Technology Life Cycles and Human Resources

Technology Life Cycles and Human Resources

Patricia M. Flynn

UNIVERSITY
PRESS OF
AMERICA

Lanham • New York • London

Copyright © 1993 by
University Press of America®, Inc.
4720 Boston Way
Lanham, Maryland 20706

3 Henrietta Street
London WC2E 8LU England

Copyright © 1988 by Ballinger Publishing Company

Library of Congress Cataloging-in-Publication Data

Flynn, Patricia M.
[Facilitating technological change]
Technology life cycles and human resources / Patricia M. Flynn.
p.   cm.
Originally published: Facilitating technological change.
Cambridge, Mass. : Ballinger, 1988.
Includes bibliographical references and index.
1. Technology—Social aspects.   I. Title.
[T14.5.F59   1992]
303.48'3—dc20   91–40843 CIP

ISBN 0–8191–8504–3 (pbk : alk. paper)

To Matthew, Eric, and Peter

CONTENTS

LIST OF FIGURES

LIST OF TABLES

PREFACE

This study grew out of two long-term research interests—how changes in the structure of labor demands affect the skill requirements of industry and how schools respond to changing skill needs. In my dissertation on the responsiveness to labor market changes of various types of educational institutions, I developed the idea that there was a training cycle in which schools played only a small role in the provision of skills that were very new and those that were fast becoming obsolete. For skills involving new technologies, on-the-job training appeared most important. For skills associated with established or declining fields, which often involved replacement needs, firms or government training programs tended to provide the necessary training. The real market for school-based skills training appeared to be for those high-volume skill needs that were established, or that were becoming established, in local labor markets.

A natural approach to pursuing these interests was examining technological change as one of the major factors altering the composition of skills at the workplace. The high-technology revolution that swept through Massachusetts in the late 1970s and early 1980s provided numerous examples of such change. Along with that revolution came widespread complaints from high-technology employers that schools were not adequately responding to their needs.

To understand the relationships among technology, firms, and schools required study at a level of disaggregation that permits analysis of the decision-making and adjustment processes of individual institutions. The Lowell, Massachusetts, area—with its fast growing high-technology sector and rapidly rising demand for labor—seemed to be an ideal laboratory for pursuing this line of research. Moreover, the high-technology–based revival of the Lowell economy has been cited worldwide as a "model for reindustrialization" for older cities and regions losing jobs in their traditional manufacturing industries.

At the same time, I was troubled by the limits of a case study of a single local labor market. Although case studies allow in-depth analysis, they are inevitably questioned as a basis for reaching general conclusions. To overcome this limitation, I compiled a more extensive data base of international case studies on technological changes and skills. Combining this data base and my case study of Lowell, I hoped to develop a comprehensive picture of the human resource implications of changing technology.

This book reports both the findings of the Lowell labor market study and the analysis of the results of approximately 200 enterprise-level case studies. The size of the case-study data base permits the identification of common patterns and generalizations as employers adjust to technological change. When viewed within the conceptual framework of production life cycles—a framework that views products, production processes, and technologies as dynamic phenomena whose skill and training requirements change as they evolve—these data help reconcile many of the inconsistencies found among previous research reports.

Facilitating Technological Change: The Human Resource Challenge develops the concept of the skills-training life cycle over the course of development of a technology. The skill-training life cycle is the human resource counterpart of the traditional product and technology life cycles. The book also organizes many of the puzzling issues of how employers and schools interact as they do and when they do, and it provides new insights into human resource planning for technological change.

Facilitating Technological Change is written for practitioners and scholars in a variety of fields. It is intended to help managers facilitate technological change at the workplace. It also seeks to elucidate various facets of production and technological change for educators striving to meet the needs of changing labor markets. For students and scholars the book provides a deeper understanding and interpretation of the vast array of research findings on the effects of technological change on skills, jobs, and workers. Finally, it is hoped that the book will contribute to the development of public and private policies for human resources that acknowledge and integrate the dynamics of the processes of technological change and economic development.

The resources to pursue this research were provided by grants from the National Institute of Education (NIE-G-82-0033) and the Employment and Training Administration of the U.S. Department of Labor (21-25-82-16). Bentley College also provided generous support through a Rauch Faculty Enrichment Fund grant, a Faculty Development Fund award, and graduate and undergraduate student research assistance. The Federal Reserve Bank of Boston provided a stimulating environment for writing the results of the Lowell study during a sabbatical leave in 1983–84.

Many people have helped make this book a reality. I am grateful to Linda Byam, Paul Davis, Adam Seitchek, Ann Spruill, Jill Dugan, and Jyoti Aggarwala, all of whom served tirelessly as my research assistants. Paul Osterman, Lynn Browne, and Constance Dunham provided helpful comments on various drafts. Hedva Sarfati, as well as the library staff at the International Labor Organization in Geneva, provided invaluable assistance in identifying and procuring case study materials. Benjamin Chinitz, Henry Przydzial, David Cordeau, and staff members of former Senator Paul Tsongas's office were instrumental in arranging for local contacts in the Lowell area. I would also like to thank all of the educators, employers, and government officials whose generous contributions of time and knowledge made the Lowell study possible. Special thanks go to Joan Donlan, Jeremy Gowing, Katherine Lynch, Susanne Nadeau, John Rees, and

Mary Vlantikus for their encouragement and support in a variety of ways as the book was progressing. I also am indebted to Marjorie Richman, Barbara Roth, and Penny Stratton, each of whom managed various aspects of developing and editing the manuscript.

Finally, I am particularly grateful to my husband, Peter B. Doeringer. Peter not only provided insightful comments and helped me sharpen my thinking throughout the research, but he was also a continuous source of encouragement and moral support.

I TECHNOLOGICAL CHANGE AND HUMAN RESOURCES

1 INTRODUCTION: TECHNOLOGY, JOBS, AND HUMAN RESOURCES

There are many popular perceptions about how technological change affects society. The vast literature on the topic shows a variety of approaches and assumptions, and substantial variation among conclusions. Amid the myriad research studies on the effects of technological change at the workplace and on local economic development one can find evidence to support almost any thesis.

TECHNOLOGICAL CHANGE AT THE WORKPLACE

With respect to skill requirements, for instance, some studies claim that technological change causes deskilling, that is, the dismantling of complex tasks into more simplified assignments.[1] Others aver that the adoption of new technologies will generate tasks requiring higher skills.[2]

The reported effects of technological change on jobs also are diverse.[3] Technological change is found to create, eliminate, enlarge, and deskill jobs. New high-skill tasks may be added to existing positions, for instance, resulting in "job enlargement." Alternatively, entirely new positions may be created. The emergence of lower-skill tasks generates new unskilled positions in some instances and leads to the deskilling or elimination of existing jobs in others. In addition, the average level of skill

3

increases after the introduction of technological changes in some firms but remains constant or decreases in others.

Net employment also has been shown alternatively to increase, decrease, or remain stable in firms that have adopted technologies.[4] Even individual firms adopting similar technologies have been found to have significantly different experiences with employment gains and losses.

Workers, too, are affected differently by technological change.[5] Some workers benefit through upgrading or promotion—and some by being newly hired by the adopting firm. Other workers are downgraded, laid off, or forced to relocate in order to remain with their employer. All these shifts in employment may even occur simultaneously within a firm.

TECHNOLOGICAL CHANGE AND LOCAL ECONOMIC DEVELOPMENT

Technological change has significantly diverse impacts not only on firms but also on communities.[6] In some areas plants close and unemployment rises when technologies are adopted.[7] In contrast, others experience rapid growth and economic revitalization. [8]

In recent years, the introduction of "high-technology" industries has been widely advocated as a means of reviving depressed economies.[9] As generally defined, high-technology industries are those operating at the "cutting edge" of new technologies. They usually are identified by their relatively high proportions of research and development expenditures and of professional and technical workers.[10]

The appropriateness of deliberately fostering structural change along high-technology lines has not been carefully scrutinized. Depending on the definition used, high-technology industries account for only a relatively small proportion of jobs in the United States—ranging from under 2 percent to approximately 13 percent.[11] Moreover, employment in high-technology industries has been found to be quite volatile.[12]

Communities, states, and nations continue to seek high technology employers actively, in spite of the ambiguity that persists over the quantity, quality, and duration of jobs that this sector can deliver. High-technology industries are characterized by significant heterogeneity with respect to products, occupations and skill requirements, wage levels, firm sizes, and organizational ownership arrangements.[13] They encompass firms operating in a variety of markets from microprocessors and space-age products to household appliances and home-movie film. Firms operating on the frontier of defense technology are included, as are those in well-established consumer businesses. Some firms in high-technology industries produce to order, others mass produce their products. Some companies manufacture final products, others provide subcontracting services. In addition, high-technology industries include firms that are large, modern, and growing rapidly, as well as firms that are small and marginal.

Policy discussions on skill needs for high technology have focused on engineering and technical jobs.[14] High-technology industries, by definition, do employ a relatively large proportion of highly skilled professional and technical positions compared to other types of industries—and yet most of the jobs in these newer industries are blue-collar and clerical, and involve lower-skill and unskilled positions. In addition, because of the wide range of skill requirements and occupations, wage levels vary considerably both among and within high-technology industries.

Debate continues over the factors that influence the location of high-technology industries. Empirical evidence shows widespread diversity of high-technology sectors across geographic areas.[15] The occupational composition varies by region, as does the distribution of firm sizes and ownership arrangements. Large-scale statistical studies on the relative impact of factors such as taxes, wage rates, and transportation costs on the spatial patterns of high-technology employment have been generally inconclusive.[16] Survey data of employers in high-technology industries also are diverse. Some studies highlight the importance of a supply of professional and technical talent and agglomeration economies derived from an established high-technology base.[17] Other studies, citing the dispersion of employment in high-technology industries to lower-cost areas, suggest that tax breaks and a supply of low-cost labor are important factors in a technology-based economic development strategy.[18]

A NEW, HIGHLY DISAGGREGATED APPROACH

Much of the confusion over the impacts of technological change stems from the fact that studies are often conducted at such high levels of aggregation that the actual effects of the changes are masked. Industry- and nationwide studies generally fail to capture the specific details and interrelated effects of technological change. Employment data often are not sufficiently detailed to show the wide range and changing nature of skills used within firms and industries. Data indicating that average skill levels have risen or fallen in a particular country or for a certain industry suggest very little about how technological change itself affects jobs and workers. Similarly, shifts in broad occupational categories reflect a wide variety of economic and social factors.[19]

The problem of using industry aggregates to study the impacts of technological change has two key dimensions. First, the skill and job impacts of technological change are fundamentally firm-level phenomena.[20] The tremendous variety of experiences of employers cannot be explained by industrywide studies. Second, aggregation across technologies and product lines, whereby technologies and products at different stages of development and maturity are viewed collectively, hinders analysis of the process of change.[21]

The most relevant and comprehensive data on the effects of technological change are found in detailed case studies on individual firms. These case studies can indicate what new skills are required; the relative change in skill requirements; the types of new jobs created; which jobs are deskilled or eliminated; how internal job hierarchies are affected; the role of labor-management agreements in affecting change; and the role of external factors in decision-making processes.

Such case studies also shed light on how workers are affected by these changes in skill mix and job structure. They illustrate how the newly created jobs are staffed; who provides training for new skill needs; what criteria govern the allocation of work tasks; which workers are upgraded, laterally transferred, or demoted after the change; which workers are laid off; and how the timing of the adoption influences a firm's adjustment processes.

Similarly, micro-level analyses of local economies and business decision making, not nationwide studies, are the key to unraveling the effects of technological change on local economic development.[22] The mix of firms and production activities in an area will influence the extent to which technological change will result in layoffs, plant closings, and unemployment. The established local employment base and labor market also will determine the conditions under which newer firms and industries will compete for workers.[23]

To date, no unifying model or approach has been devised to reconcile the conflicting findings, nor to provide guidance in facilitating technological changes at the workplace and in the local economy. Studies of technological change at the level of individual firms are usually descriptive and lack a theoretical underpinning. As a result, they are often viewed as "special cases," isolated from other examples and lying outside the context of processes of change more generally. Such case studies, therefore, tend to lack generality in helping to explain the impacts of technological change on a larger scale.

This book, drawing upon both a large body of case materials and original field research, demonstrates that production life-cycle models provide a framework to systematically analyze the effects of technological change. This framework—in which products, production processes, and technologies are seen as dynamic phenomena whose skill and training requirements change as they evolve—provides a theoretical model from which to draw generalizations and common themes.

The use of detailed micro-case studies limits quantitative assessments and forecasts of the effects of technological change, such as the overall extent of deskilling, upgrading, transfers, and the like. Nevertheless, this level of disaggregation permits an in-depth view and understanding of technological change and adjustment processes not otherwise possible. It also allows the factors responsible for divergent outcomes from technological change to be identified.

By viewing technologies and products as evolving phenomena with changing skill requirements, the life-cycle framework helps to reconcile many of the inconsistencies in the previous findings. Moreover, by providing a detailed understanding of the dynamic processes of technological change and its effects on skills, jobs, and training, it casts a new perspective for business and public policy.

The following chapter presents the life cycle models for products, processes and technologies, and extends these traditional models to an analysis of changing skill and training requirements. In this framework, Part II assesses empirical evidence from approximately two-hundred firm-level case studies on the effects of technological change at the workplace. It demonstrates patterns in skill requirements, jobs, training needs, employer staffing practices, and worker career paths over the life cycle of a technology.

Part III focuses on the impacts of technological change in the community, using an in-depth case study of a local economy transformed from a depressed, stagnating area into a booming center of high-technology employment. The study analyzes the technology-induced changes in skill requirements and training needs during this transition, highlighting the skills-training cycle that occurs as the provision of job skills shifts from the workplace to the schools as technologies mature. The conclusions and implications for human resource policies to facilitate technological change at the workplace and in the community are presented in the final chapter.

2 PRODUCTION LIFE CYCLES AND HUMAN RESOURCES

Analyses of the effects of technological change on skills and employment are usually cast in a static framework that involves a series of discrete jumps from one technology to another — each with unique skill needs. Life-cycle models, in contrast, postulate sequential development paths — from birth to growth, maturity, and eventually stability or decline. Along these paths, technologies, production scale, and work organization evolve, affecting skill needs in complex ways. These interrelated changes occur in predictable fashion, however; the timing and patterns of the different phases are determined by variables such as risk, level of demand, and degree of standardization of products and equipment.

Product, process, and technology life-cycle models, originally formulated for devising marketing and sales strategies, can offer new insights into understanding the effects of technological change at the workplace and in local labor markets. This chapter reviews these life-cycle models and extends them to skill requirements, job structures, and training needs. A "skill-training life cycle" is postulated, showing the relationships among changes in skills, providers of training, and the impact on job structures as technologies mature. The implications of these models for human resource planning for technological change are then addressed.

PRODUCT, PROCESS, AND TECHNOLOGY LIFE-CYCLE MODELS

The Product Life Cycle

The "product life cycle," a term coined in 1950, depicts phases of development of an individual product.[1] The most common version of the product cycle portrays four stages of sales growth: introduction, rapid growth, diminished growth, and stability or decline. The introductory stage of the cycle is characterized by research and development (R&D) and innovation. New products are introduced and manufactured with considerable variability in design and quality.[2] The second stage of product development involves rapid progress and growth. Following an initial period of experimentation, products are improved, quality becomes more consistent, scale economies permit lower costs, and demand rapidly accelerates. Price competition begins, and greater emphasis is placed on advertising and on the development of new markets.

Maturity characterizes the third stage of the product cycle. Product demand continues to expand, but at a slower pace, and replacement demands become an increasing share of product sales. Price competition intensifies, imports threaten domestic production, and the marketing emphasis shifts to product differentiation. In the final stage of the product life cycle, sales level off or fall. Products face obsolescence as consumer demands shift to new, substitute products.

In its simplest form, the product life-cycle model suggests one development pattern for all products, but more sophisticated analyses indicate a variety of possible growth paths.[3] This variation, however, is not a random phenomenon. In particular, the timing and shape of a product's development stages are determined by a series of factors, including the nature of the product, the rate of market acceptance, ease of entry into the industry, and the rate of technological change. Some products skip a phase of development. Others never evolve from small-batch, customized production to highly standardized goods produced in large volume. Product improvements and the tapping of new markets can extend a product's "life," as can the absence of new product substitutes. In addition, products may "recycle" through the development pattern in response to new innovations or the discovery of alternative uses of the product. Moreover, the evolutionary cycle of an individual product is malleable. A firm's marketing and innovation strategies can influence the course and timing of a product's development.

The Process Life Cycle

Coincident with changes in the nature of and the demand for the product are changes in production processes.[4] More specifically, standardization and large volume foster the exploitation of economies of scale in production. (See Figure 2–1.) Flexible, trial-and-error methods of production abound in the initial stage of a product's development. Job-shop, or relatively short, production runs are used during this period as considerable uncertainty characterizes product design and demand. Much of the work is initially done by hand. With the adoption of general-purpose equipment during the relatively early phases of product development, major changes in the product can still be easily accommodated.

With standardization of the product, the locus of change and innovation shifts from product to production process. Product assembly becomes increasingly more automated. In addition, equipment becomes more specialized, making further product changes relatively costly. Capital-intensive mass-production techniques replace small-batch production as products mature and as competitive advantage increasingly becomes a function of cost minimization.

Product and Process Life Cycles and Employment

The implications of product and process life-cycle models for spatial employment patterns are two-dimensional. Patterns of regional specialization of employment should occur as employers seek to locate different production activities in areas best

Figure 2–1   Product Life Cycles.

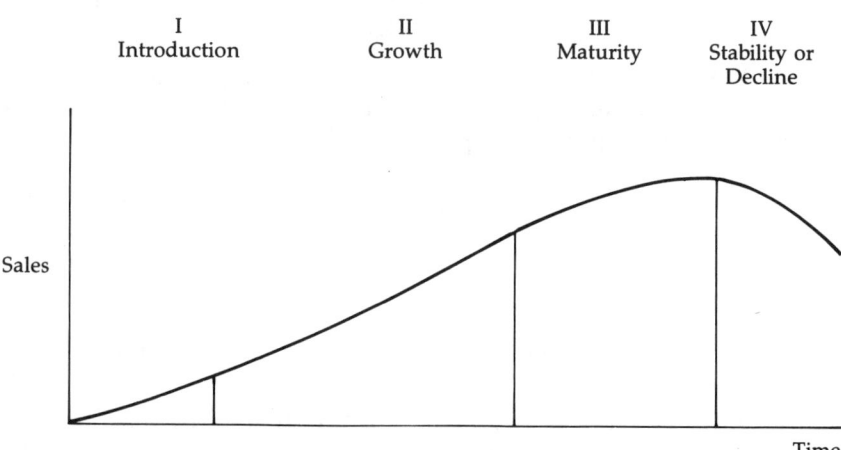

Characteristic	Phase			
	I	**II**	**III**	**IV**
Product	Variable; often custom designs	Increasing standardization		Mostly undifferentiated; standardized
Product innovation	Frequent experimentation; major changes	Declining rate		Minor refinements, if any
Volume	Small-scale	Rising volume	Large scale	Large scale
Process	Job-shop; batch production	Increasingly automated	Capital-intensive mass production	Capital-intensive mass production
Process innovation	Exploratory	Relatively high rate; major innovations	Rate declines	Minor refinements, if any
Equipment	General purpose	Increasingly specialized		Special purpose

Adapted from: William J. Abernathy and James M. Utterback, "A General Model," Chapter 4 in W. J. Abernathy, *The Productivity Dilemma* (Baltimore: Johns Hopkins Press, 1978), pp. 68–84.

suited to their needs. In addition, changes in the mix of production activities over the life of a product should trigger shifts in employment.

Extensions of the product life-cycle model, in particular, explain the nature of these trends. The "international product cycle model," for example, suggests that firms initially locate close to the source of demand for their newly developed products so they can rapidly communicate market information into product changes.[5] When foreign markets create demands for the product, they initially generate exports for the producing country. At some point, depending on the nature of the products and the characteristics of foreign demand, the expanded foreign market attracts its own production base.[6] When production costs abroad are low enough to compensate for transportation and other costs, such as tariffs, the country that originally produced the product becomes a net importer of the good. At final stages of product development, production activities may shift from the sites of product demand to lower-cost areas in other countries.[7]

The international product cycle model suggests that countries specialize in producing goods consistent with their comparative advantage.[8] Industrialized, capitalist countries will tend to focus on their research and technical capabilities critical in new, rapidly growing industries, whereas less developed countries will take advantage of their relatively low labor costs.

The "regional life-cycle model" suggests similar spatial production patterns for smaller geographic areas.[9] With respect to human resources, in particular, the model implies that the attractiveness of regional and local economies varies with the skill needs of products at different stages of development. Early stages of product innovation and development are said to occur in areas in which highly skilled professional and technical workers are available to conduct R&D. Standardization and increasing output of the product trigger reduced labor requirements, inducing production shifts to geographic areas characterized by lower labor costs.

Empirical evidence supports the spatial employment patterns of these international and regional models. Industries usually rely on a range of technologies and have products in several phases of development. They therefore engage in a mix of production activities characterized by diverse skill needs and employment patterns. Innovations, R&D, and new product activities, for example, require highly skilled workers, tend to be highly concentrated geographically, and are relatively stable in terms of location.[10] As long as new models and major design changes are being introduced quite regularly, employers will not want to separate the design and testing functions from product assembly. At later stages of development, more standardized products and equipment permit the separation of production from R&D. These relatively routinized assembly functions of mature products require low-skilled labor, and they generally disperse to geographic areas characterized by relatively low production costs.[11]

The electronics industry, for example, produces both highly sophisticated products that incorporate technologies on the cutting edge as well as more mature consumer electronics goods, such as radios and televisions. Firms manufacturing the newer goods tend to concentrate their production operations near R&D. More mature products are produced in lower-cost areas.[12] Similarly, while an increasing share of the world supply of semiconductors is produced outside the United States, in countries with relatively abundant supplies of low-cost labor, the design and development work is still highly concentrated in Silicon Valley.[13]

The computer industry shows similar patterns of regional specialization and employment trends. R&D, design, and production of state-of-the-art equipment continue to be quite concentrated. In contrast, the large-scale production of relatively standardized computer components and routinized assembly activities are dispersing away from the R&D centers. In 1979, for example, two-thirds of the computer industry in the United States was concentrated in two sites: New England and California.[14] Production of peripheral equipment, however, was quite decentralized. Printers and terminals generally were manufactured in large branch plants in states with relatively low labor costs such as Tennessee, South Dakota, and North and South Carolina. Individual computer components are often produced in low-wage countries such as Mexico, Hong Kong, and Taiwan. In the minicomputer segment of the industry, design and initial assembly remains geographically concentrated in Massachusetts and in California, along with company headquarters. The production of the more standardized components and peripheral equipment for minicomputers, however, has dispersed to other states, for example, North Carolina and Arizona.[15]

Industries with relatively little large-scale production, such as manufacturers of medical instruments, customized electronics equipment, and communications equipment, also regionalize their operations—but to a lesser extent.[16]

The Technology Life Cycle

Changes in technologies are often the stimulus behind product life cycles and modifications in production processes. Moreover, individual technologies—such as a numeric control technology, a microelectronics technology, or a data-processing technology—have life cycles of their own.[17] A technology, introduced slowly at first, becomes more widely adopted as intensive R&D efforts lead to improved performance. Eventually it reaches peak performance—and is often later replaced by a new, superior technology.

There is no one-to-one relationship among technologies and products and production processes—several product and process life cycles may evolve with the development of any one technology. Technological evolution, however, can often signal impending changes in products and in production processes.[18] For instance, as a technology matures, uncertainty about its capabilities and limitations declines, and products and production processes can become more standardized. Rapid product innovation accompanies the earliest phases of a technology's developments, whereas process innovation peaks later in the technology's cycle as product design stabilizes. Innovations in the later stages of development of a technology, when they occur at all, are primarily minor improvements in equipment rather than major or fundamental changes in either product or production process.

Other aspects of technological change can also affect the timing and shape of product and process cycles.[19] Rapid technological change, for instance, increases uncertainty and hinders the move toward standardization. By accelerating product obsolescence, rapid technological change also shortens the effective "lives" of products.

The rate of diffusion of technologies influences the development of products and production processes as well. In general, diffusion rates are said to be slower the more complex the technology, the more specialized the technical personnel required, the heavier the capital requirements, and the greater the economies of scale in R&D.[20] Patents and the continual development of new technologies hinder the diffusion process. Yet as technologies — even those "protected" by patent — mature, knowledge about them becomes more widespread as a result of personnel turnover, physical inspection of final products by competitors, and the like.

THE SKILL-TRAINING LIFE CYCLE

A skill-training life cycle evolves as the level of demand and standardization of skills changes with the development of a technology. The early stages of a technology, characterized by a high degree of product innovation, are relatively skill- and labor-intensive.[21] Engineers and scientists are needed to develop new products, construct pilot models, and implement design changes. These professionals also perform most of the tasks later assumed by production and marketing managers, technicians, and skilled craftsworkers. The general-purpose equipment that characterizes the earlier stages of product development requires skilled operatives able to adapt the equipment to the individual company's needs. These operatives are required to perform a wide range of tasks and to adjust to frequent changes.

As technologies mature, standardization and the expanded use and complexity of equipment permit a greater division of labor and the subdivision of multifaceted tasks into more narrowly defined assignments.[22] As tasks become "deskilled", the workers' skills, experience, and independent decision making become less important.[23] The tasks of semiskilled operatives, for example, often shift to monitoring and control of the equipment. In addition, product assembly can be done by low-skilled and unskilled workers who concentrate on a very limited number of specific tasks. Once embodied in the work force, skills are transferred to the production equipment.

The effects of technological change on skill requirements at the workplace depend on the timing of the change relative to the stage of development of the technology. More specifically, the technology life cycle suggests that the development and introduction of new technologies generate relatively high-skill professional and technical needs. As a technology matures, some relatively high-skill maintenance and repair tasks also can be expected. Concurrently, increasing levels of standardization and mass-production techniques cause the deskilling of a wide range of tasks. In its extreme form, the deskilling process results in the elimination, rather than just the simplification, of certain tasks.

The focus of deskilling shifts over the technology life cycle. Generally, the skill level of the task being simplified is inversely related to the degree of standardization of the products and production processes. When equipment is initially introduced into small-batch production, high-skill handicraft work such as that of machinists and welders is simplified or eliminated. The automation of routinized assembly functions, in contrast, eliminates relatively unskilled tasks.

Moreover, the maturation of a technology can cause the deskilling of tasks that were created earlier by that same technology.[24] The need for skills required for setting up general-purpose equipment in factories, for instance, can be eliminated when more sophisticated, specialized equipment is adopted. Similarly, data-entry keypunch tasks created by computer adoptions can be eliminated with the installation of more complex computer equipment.

The availability of skill training and the mix of institutional providers vary as job skills undergo phases of introduction, growth, maturity, and stability or decline.[25] When a technology is new, skill training is usually provided on the job or in programs at the workplace.[26] In the earliest stage of a technology, scientific and engineering personnel design and create a variety of experimental products. Subsequently, they teach others what needs to be done for small-batch production. (See Table 2–1.)

After a technology becomes more widely adopted and equipment standardized, skills that were once "firm-specific" become "general" skills that are transferable among employers.[27] As with products, increased demand and the standardization of skills permits their "production" on a larger scale and at locations away from the R&D sites. As employers cannot capture the return on investments in general skills, they prefer to shift such training out of the workplace and into the schools, where the government or individual students will pay for it. Moreover, as demand for such skills grows, it is easier to formalize the training and to provide it in the schools. Together, these two forces encourage the shift of skill development from the workplace to the formal educational system as technologies mature. Computer programming, keypunching, word processor training, and the setup and operation of numerical control equipment are classic examples of this transfer.

Schools and colleges differ widely in their missions, funding arrangements, and decision-making procedures.[28] As a result, types of training are diverse, as are the manner and time frame in which the institutions respond to labor market changes. Initially, training is offered by schools and colleges that are oriented toward meeting the needs of employers. As demands for the skill mature, training becomes more widely diffused among educational and training institutions.

After a technology "peaks" and the demand for a skill declines, training becomes less available at the various educational and training institutions. The pattern of elimination and contraction of skills programs again reflects the priorities and constraints of educational institutions. Once a technology becomes obsolete, the locus of skill training may revert to the firm as it seeks to fill its relatively short-term, skilled replacement needs.

Given the nature of skill emergence and development, mismatches in the demand and supply of skills often occur as a technology evolves. In particular, in the early phases of the skill-training life cycle, skill shortages appear due to implementation and training lags in the educational network. Alternatively, surpluses of newly trained graduates can arise if providers

Table 2–1   The Skill-Training Life Cycle.

	Phase			
	I Introduction	II Growth	III Maturity	IV Stability or Decline
Tasks	Complex	Increasingly routinized		Segmented
Job skills	Firm-specific	Increasingly general		General
Skill training provider	Employer or equipment manufacturer	Market-sensitive schools and colleges	Schools and colleges more generally	Schools and colleges; some skills provided by employers
Impact on job structures	Job enlargement; new positions	Emergence of new occupations		Rigid job hierarchy with formal education and occupational work experience requirements.

of job-related skills do not accurately predict patterns of declining demands and of skill obsolescence as technologies evolve.

Technological Change at the Workplace

The skill-training life cycle affects the impacts of technological change at the workplace as it alters employers' hiring and staffing patterns and workers' career paths.

As a technology matures, the emergence of new occupations diminishes the likelihood of job "enlargement." The unsettled environment caused by the early adoption of a new technology encourages firms to incorporate the newly created tasks into existing jobs rather than add new types of positions to job hierarchies. Adopters of newly developed technologies face considerable ambiguity regarding the quantity and level of skilled personnel required to accomplish new tasks. Moreover, as long as production processes continue to undergo substantive changes, firms face considerable uncertainty regarding industry standards and new generations of equipment. Relatively high equipment costs encourage firms to lease rather than purchase equipment, further highlighting the need for firms to maintain flexibility in staffing.

In contrast, firms that adopt relatively mature technologies face fewer of these uncertainties, for specific occupations have emerged in those fields. As noted earlier, the increased levels of standardization and mass production permitted by the maturation of technologies foster the fragmentation of relatively complex tasks into more routine, less skilled assignments. Coupled with growing demands for these skills this increasing division of labor promotes the creation of a more well-defined set of occupations. Relatively late adopters of data-processing technologies, for instance, know that they need to fill a variety of computer-related positions, including computer systems analyst, computer programmer, computer technician, and computer operator. In contrast, earlier adopters depended on a more amorphous, broadly trained group of workers to fill new skill needs.

Technological maturation is likely to disrupt existing career ladders or "internal labor markets" within firms.[29] Employers frequently recruit new personnel to fill positions relatively low in the job hierarchy and promote from within.[30] Hence many of the better-paying, more high-skill jobs are sheltered from the external labor market. Internal labor markets work to the advantage of both employers and workers: the employer is given the chance to assess the individual's performance prior to his or her promotion into a job requiring greater responsibility, skill, and independent decision making. Internal labor markets also provide career incentives and advancement opportunities for workers. These in turn enhance worker commitment to the firm and can reduce costly turnover.

The skill-training life cycle suggests that as technologies mature, a growing proportion of the relatively high-skill tasks created at the firm will be filled by workers from the external labor market. In the early stages of development of a technology, the relative importance of "job enlargement," fostered by the high degree of uncertainty and a lack of appropriately trained workers, favors the selection and training of current employees to perform many of the new tasks. At later stages, in contrast, current workers are more likely to be bypassed as individuals trained outside the firm are hired to fill the better jobs created by the technological change. Moreover, as schools and colleges take over the training responsibilities in these fields, educational credentials become associated with particular skills and occupations. Previously established internal job ladders are jeopardized as formal educational requirements or related occupational work become criteria for certain skilled positions.

The skill-training life cycle suggests that the impact of technological change at a particular worksite will reflect a variety of firm-specific characteristics. For one thing, management decisions on the timing of adoptions, relative to the development of the technology as well as to economic growth and product demand, influence the effects of the technological change at the workplace. In addition, whereas technological change directly affects skill requirements to accomplish various tasks, the impact of such change on specific *jobs* depends on how those tasks are integrated into jobs.[31]

Management practices and labor-management agreements come into play here, influencing not only job changes but also the allocation of those jobs among workers. Reliance on job security provisions or departmental seniority to staff new positions, for example, may enable displaced workers to be transferred into relatively high-skill jobs. In the absence of such arrangements, workers other than those displaced may be selected for these positions.

The effects of technology-induced changes in skill requirements also will differ by firm size. Due to relatively limited training budgets small- and medium-sized firms are likely to be much more dependent on external sources—such as schools and colleges, government training programs, and other firms—for their skill needs.[32] Smaller firms are at a particular disadvantage with the adoption of new technologies, as employers are relied on to provide training in new and emerging fields. They may also find it harder to retain skilled workers during times of skill shortages as larger firms are often able to offer higher wages and greater promotion opportunities.

Technological Change and the Community

Technological change can generate a series of new, relatively high-skilled jobs and thereby contribute to the economic growth and development of an area. Alternatively, technological change can trigger the relocation to other geographical areas of low- and unskilled jobs involved with relatively standardized production processes.

The mix of firms in an area influences the community's vulnerability to the destabilizing effects of technological change. Firms with products and technologies at various stages of development have different skill and training needs. Communities offer a wide range of human resources and amenities, and hence are likely to exhibit considerably different employment bases and skill needs.

Production life-cycle models accentuate the importance of employment diversity in an area for long-term economic development. A diversified employment base mitigates the likelihood of significant swings in an area's economic activity as technologies evolve. The local mix of jobs affects the availability of alternative employment opportunities to cushion the impact of technology-induced plant closings, layoffs, and unemployment. The local employment base also influences the competitiveness of the area in attracting new, emerging employment opportunities.

SUMMARY

Extensions of the traditional life-cycle models suggest that employers and local communities should expect to experience ongoing changes in the nature of their skill, training, and employment needs as technologies evolve. A "skill-training life cycle" is proposed relating changes in skills with the availability and provision of skill training. This cycle influences the impact of technological change at the workplace by altering employers' hiring and staffing practices as well as workers' career paths. Skill shortages and surpluses, job creation and deletion, and job relocation also are seen as "natural" outcomes of technological change, with significant implications for local economic development.

The production life-cycle framework highlights the relationship among these dynamic trends over a technology's life cycle. These patterns will affect the impact of technological change in particular workplaces and communities, and it can be used to guide public and private policies to facilitate such change. The following chapters provide empirical evidence of the conceptual framework presented. Thereafter the human resource policy implications of the models are discussed.

II TECHNOLOGICAL CHANGE AT THE WORKPLACE

3 TECHNOLOGICAL CHANGE, SKILLS, AND JOBS

By influencing the types of skills and training needs created by technological changes as well as the availability of appropriately-trained workers, technology life cycles critically influence the ways firms integrate changing skill requirements into jobs. This chapter takes a closer look at the relationships between technological change, skill requirements, and jobs. Empirical evidence on the effects of technological change is provided from a data base of approximately two hundred case studies of firms that undertook technological adoptions.

THE DATA BASE: A BRIEF OVERVIEW OF THE CASE STUDIES

The data base, comprising 197 case studies published between 1940 and the present, was compiled specifically to investigate the impacts of technological change at the workplace. The cases span a wide variety of industrial organizations, technologies, and economic and social circumstances.[1] The individual firms studied vary in size and include both public and private enterprises, manufacturing and nonmanufacturing establishments, and unionized and nonunionized companies. Approximately half the cases refer to automation in manufacturing; the other half pertain to office automation.

The cases are international in scope: over a dozen countries are represented. Firms in the United States account for just over half the sample. England contributes the second largest number of cases, followed by West Germany. Other countries represented include France, Canada, Sweden, India, Australia, Japan, Russia, Norway, Wales, and Austria. (See Appendix 3A for additional information on the case study data base. Also see the Annotated Bibliography of the Case Study data base at the end of the book.)

Each case study discusses the introduction of new equipment or control devices associated with a technology new to the particular firm. A range of different technologies — such as programmable automation, microelectronics, data processing, and word processing — is involved. In addition, the timing of the adoptions varies relative to the technology life cycle. In many cases, while "new" to the firm, the technology is in later stages of its development when adopted. As we shall see, this distinction is vital to the understanding of the differential impacts of technological change on skills and jobs.

TECHNOLOGICAL CHANGE AND SKILL REQUIREMENTS

Evidence from the case studies demonstrates that technological change at the workplace both creates new high-skill requirements and deskills existing tasks. These results occur when technological change affects work that is highly skilled as well as when relatively low-skilled jobs are affected. They also occur whether the change takes place in a factory or in an office.

The Creation of Relatively High-Skill Tasks

Manufacturing

The adoption of technologies in manufacturing has led to increased skill requirements needed to perform a variety of tasks involved with machine operation and setup, maintenance and repair, computer programming and systems analysis, and equipment design.

When machines are installed to perform production tasks that were previously manual, more highly skilled requirements are usually created. The automation of low-skilled manufacturing tasks — such as the introduction of automatic pressers in apparel firms and of numerically controlled wiring equipment in the manufacture of electrical equipment — creates demands for a higher level of skills to operate the machines and to adjust regulatory devices. For example, one case study shows that operatives needed greater skills when automated machinery was introduced in a fabricated metals firm:

> An operator of the newer machine must be of a higher skill level, must have more mechanical knowledge and comprehension than any worker under the previous production arrangement. He exercises judgment in knowing when dies require adjustment or need to be replaced, must lubricate and tend the machine, be able to stop it in case of a jam, cut out and remove the jammed strip, refeed the steel strip through the machine, and resume operations. The operator is not required to change the set-up of the machine.[2]

Cases on automatic assembly equipment, installed in factories in the 1940s and 1950s, also illustrate the higher level of skills required for setup activity. In an electrical machinery company, for example:

> Machine preparation and adjustments are more complex, more difficult, and require a wider range of technical knowledge and competence because of equipment complexity and the intermingling of hydraulics, pneumatics, electrical and electronic, and mechanical adjustments. Also workfeeding devices must be synchronized and otherwise regulated and new degrees of precision must be obtained by careful adjustment.[3]

In addition, more advanced skills are usually required for maintenance functions once relatively sophisticated equipment is installed. With the introduction of route relay control and the dieselization of railways, for example:

> On the maintenance side [with dieselization], a whole series of new skills have now been brought into being to maintain the diesel locomotive or sophisticated signalling equipment. These are jobs which will require a much greater technical knowledge, skill and concentration. For the categories of maintenance of the new equipment new skilled categories of

electric signal maintainers and mechanical signal maintainers will be now be required and shortages in these professions will experienced. These employees will be required to bring to their "profession" a knowledge of the maintenance of electro-magnetic and electronic equipment and will, therefore, be equated to a higher scale.[4]

The case studies demonstrate that the introduction of electronic control systems in manufacturing also creates a need for greater skills for maintaining the equipment, whether highly skilled craftswork or lesser- skilled production work is involved. Plant electricians, in particular, usually are considered incapable of maintaining such equipment without further knowledge and training.[5]

Similarly, the introduction of advanced instrumental and control devices in continuous-process manufacturing operations creates relatively skilled tasks. When computers are installed in foundries, chemical plants, and petroleum refineries to monitor and adjust production conditions, advanced technical knowledge is required to perform supervisory and maintenance tasks. Expertise in electronics, in particular, is needed to supplement more traditional mechanical and electrical skills in maintaining and repairing the new systems.

Computerized process control systems installed in a variety of manufacturing industries also generate new high-skill requirements in computer programming and systems analysis. When such a system was introduced in a steelworks, in the early 1960s, for example:

> The essential requirement of the new steelmaking and rolling department was for staff who were more technically and scientifically qualified than previously .... Automated equipment, computer control planning and information handling, all new to the Company, required the engagement of specialist staff who proved to be difficult to recruit.[6]

In firms that adapt production equipment to meet their specialized needs, new engineering skills are needed to redesign the system and get it up and running. As explained in a case study of a U.S. electrical machine company:

> A highly specialized "homemade" machine is serviced by the equipment development group that built it until it is regarded as more or less trouble-free.[7]

Offices

Office automation also generates relatively high-skill needs for a range of tasks. Analysis of the case studies on office technologies, especially those involving data processing, portrays the changing nature of the highly skilled tasks created by the adoptions. Early adopters of data-processing technologies needed highly skilled workers for computer programming and systems analysis. An insurance company adopting electronic data processing in the late 1950s, for instance, needed people who could:

> shape the office's systems to fit the machines, bearing in mind the form in which raw data first becomes available for processing and the ultimate end products which this data must be used to turn out. . . .

> translate the requirements of the analysts into a language on which the machine can operate so as to produce the desired results in the most efficient manner . . . .

> understand the basic principles of the machine's operation and . . . handle the machine and its peripheral equipment in a careful and methodological fashion.[8]

The installation of an automatic reservation system in a large airline reservation office, in the 1950s, created high-skill tasks in electronic data-processing research.

> This group is comprised of five "systems engineers." These professionally trained persons perform duties which involve planning systems development and extending electronic methods to all clerical activities of the company. . . . The qualifications for systems engineers include education at the college level and cover a variety of airline experience.[9]

In addition, the cases pertaining to adoptions of data processing in the 1950s document demands for engineering skills to perform maintenance and repair functions. In contrast, firms adopting data-processing systems years later needed new technical—rather than engineering—skills for maintenance and repair tasks.

Word-processing technologies also have created some relatively high-skill tasks. Speed and accuracy have become less critical requirements in this area; however, the need to learn codes and editing, the increased responsibility associated with operating more expensive equipment, and the possibility of altering others' files have tended to raise skill needs above those required for more traditional typewriters.

The Deskilling of Tasks

The case studies provide empirical support for the "deskilling hypothesis," which holds that many tasks become simplified or redundant when firms introduce technological changes. Tasks at all levels of the skill spectrum — including professional and technical, craft, maintenance and repair, clerical, operative, and laborer tasks — are shown to be vulnerable to the deskilling process.

Manufacturing

The case studies on manufacturing show that deskilling occurs when machinery is initially introduced into small-batch production, as well as at later stages, when more specialized equipment handles larger volumes of output. Engineering tasks are deskilled in the early stages of a technology's development as technical tasks replace what was previously considered professional work. In addition, when mass-production techniques replace small-batch production, the work of machinists, grinders, blacksmiths, and the like is prone to deskilling. In many instances, the demand for these skilled crafts is totally eliminated.

According to a case study on a small optical firm's introduction of computers, the deskilling of the craftsworkers' tasks was an "unfortunate side-effect" of the firm's policy of buying the best and most modern production machinery:

> New production machinery has, for example, been installed in the surfacing department, which makes lenses on machines which grind "blank" lenses to the shape and size prescribed by the optician. . . .
>
> Before, a skilled worker would take the prescription (every lens is unique) and from this would make "rule-of-thumb" guesses as to which blank to

use, and which tools and settings to apply. . . . Blank selection, tool selection, and machine settings can now, however, be *predetermined* with the computer, which is programmed to translate the information on prescriptions into which machine settings, etc., to use. [italics in original] [10]

Prior to this change, an apprenticeship program and years of on-the-job training were necessary to master the surfacers' craft, which involved working with a wide range of precise measuring tools. After the change, individuals "carrying out instructions on a ticket which comes from a computer" could master the necessary skills in a few months.[11]

Technological change at this optical firm also resulted in the "rational segregation" of highly skilled craftswork in the glazing section into semiskilled machine-operative tasks.

In this section, the firm introduced a variety of automatic equipment, including edging machines (which shape the lens to fit the frame and put on the "V" allowing the lens to be held by the frame) and electronic focimeters (which make the tasks of lining up the lines for an accurate fit far simpler). A small production line has been set up in the glazing section, and the various tasks are rationally segregated. Taken together, the new machinery and the division of labour on production line principles take a good deal of skill out of the work.[12]

Similarly, the work of all-round machinists capable of using a variety of grinders, cutting torches, and welding machines was eliminated with the introduction of new production equipment in a modernized steelworks.[13] Numerically controlled machinery installed in an electronic testing equipment plant also eliminated the need for "highly skilled, hard-to-recruit machinists."[14] In addition, the introduction of a variety of machining and inspection equipment in the railways brought about a sizable reduction in "the older manual skills and in crafts such as boilermakers, woodworkers, drillers, riveters, platers and blacksmiths."[15]

The automation of a manually operated production plant highlights the simplification of multifaceted tasks into more narrowly defined assignments as "over-specialised operatives" replaced skilled fitters:

In the former work organisation, one or two general fitters were in charge of the manually-operated plant. The engineering staff considered that seven or eight fitters were required for the automatic machine: their work was both more exacting and less absorbing. In the first place, it was more exacting from the point of view of tolerances ("working to a micron"), which is relatively insignificant ("there are not seventy-seven different ways of fitting, only one), and then—above all—from the point of view of physical endurance, because of the need for frequent turning round the machine. At the same time, the job has been simplified to the extent that it is generally regarded as less interesting than the former all-round job ("it doesn't need real fitters"), in spite of the responsibility involved. . . .

The problem then arose of how to grade the new "over-specialised" operatives assigned to the new machine. They would be regarded as semi-skilled workers because of their level of training, although they themselves would claim to be fitters (skilled workers).[16]

Manual operative skills also became redundant with the installation of computerized control devices into continuous manufacturing processes. Triggered by changes in temperature, pressure, or quality of inputs, the new devices can identify malfunctions in equipment or in the flow of materials, and can alert operators to the need for adjustment. The more sophisticated equipment is self-adjusting.[17]

The ability to react quickly and correctly becomes a critical job requirement as operative tasks are oriented more toward monitoring and control than toward direct operation of the machinery. Whereas the level of responsibility often increases as relatively expensive equipment and higher levels of output are at stake, by and large these operative tasks become easier.

Modern equipment in a new chemical plant, for example, simplified many operative positions, for computer-generated job cards provided easy-to-follow instructions for particular processes.[18] Similarly, with the installation of computerized process controls in petroleum firms, inexperienced operators were able to operate the computerized system once they had about two days of formal training.[19]

Skilled tasks required for setting up equipment in firms across a wide range of industries also are preempted by integrated circuitry, introduced with microprocessor controls. Moreover, the advent of plug-in components has brought about the deskilling of maintenance and repair functions. With the introduction of microprocessor-controlled papercutting machinery, for example:

The integrated circuitry takes away all the skill needed in setting up the machine, so that "the newest employee becomes a skilled operator within minutes." The older model took 33 setting up operations before it could run — the "Ultima" [the new paper-cutting machine with microprocessor control systems] has merely to be turned on, the paper cut line placed against a fixed cursor, the "set" button pressed for a trial run, and then the start button pressed.... All the new range of cutters have a minimum number of moving parts, plus easily replaceable parts which minimise maintenance and virtually eliminated servicing.[20]

The introduction of microprocessor controls into sugar production shows a similar pattern:

The MPC 80 used in sugar production controls temperature levels, flows, pressures and the optimum size of the sugar crystals. . . . The new control system is designed to remove human error from routine control operations, thereby improving reliability and output. Maintenance has been simplified in the new control systems, the use of modules allowing plant control systems to be serviced by plugging in replacement modules.[21]

Automation of lesser-skilled operative and assembly functions also simplifies work tasks. The deskilling of production activities, such as coil winding and gage frame assembly work in a large electrical machinery firm, is representative of a more widespread trend that began in the 1950s, when machines assumed tasks previously performed by factory workers.

The machine mechanically performs many of the high-speed, dexterity-requiring operations. . . . [It] does much of the physical activity of transporting, work-orienting, material feeding action. . . . The operator's functions tend to be limited more and more to (a) loading the machine, (b) monitoring its action, [and] (c) performing some relatively short bit of the operation sequence.[22]

Technological change has also facilitated or eliminated low-skill or unskilled production work, such as manual lifting and handling, and tedious, repetitious operative tasks. For example, a totally automated production technique encompassing control and testing in an auto parts firm completely eliminated hard, unskilled labor.[23] Mechanization of a frozen-food plant virtually eliminated all lifting of bagged materials, shoveling, and other manual transportation tasks.[24] Adoption of a completely automated stone-crushing process deleted the tasks of crushers, graders, washers, truckdrivers, and power shovel operators.[25] Similarly, many of the hand labor and truck-driving functions were eliminated when the Wilcox continuous mining process was introduced.[26]

Further examples of technological changes that eliminated routine operative tasks include the conversion from hand scissors to machine trimmers in a ladies' garment-manufacturing firm; the installation of an automatic presser in a ladies' slip-manufacturing firm; and the replacement of manual wirers with numerically controlled equipment in an electronic equipment manufacturing firm.[27]

Offices

The deskilling of tasks is also evident in the case studies involving technological change in offices. The adoption of word processing systems usually simplifies clerical tasks, including those pertaining to the speed and accuracy of typing skills. Early adopters of data processing were able to transfer relatively routine office tasks — such as posting, card sorting, simple calculations, and the handling of documents and data — to the new equipment. Computers installed in the 1950s also took over many accounting and bookkeeping functions, such as payroll and accounts receivable, leaving tasks performed by clerical workers oriented more toward machine monitoring. As one case study states:

> Some of the jobs that were eliminated were taken over directly by the IBM 705. Many of these were the more routine, tedious tasks, requiring employees with relatively low skill; others, however, were of a higher level involving minor decision making.[28]

When electronic, programmable computers replaced their mechanical predecessors, they absorbed additional decision-making functions.

> Besides property tax billing and collection systems, other computer applications in the Department of Finance now include nearly every operation which was formerly highly labour intensive. The list of uses is very long, and includes such things as payrolls, accounts payable and receivable, general ledger, water and sewer accounts, pension funds, homeowners' loans, and investment analysis.[29]

Similarly, a recent case study on the use of informatics — that is, the automation of information — in an insurance firm discusses the widespread transfer of clerical tasks from workers to machines:

> Another major and largely direct effect [of informatics] has been to reduce the labour content of insurance work, with most of the shrinkage in the clerical component of that work. Examples range from the partial automation of word-processor-generated policy documents and promotional materials to the more recent moves toward automatic underwriting. Here, routine issue writing, which could be construed to represent the clerical component of the underwriter's job, plus traditional clerical support functions, are being transformed from labour-intensive to capital-intensive operations.[30]

Secretarial work, more generally, has changed with the introduction of computers into offices. Many of the case studies suggest that as machines assume various office functions — such as layout, correct hyphenation, and the need for typing accuracy — fewer skills are needed.[31]

Case studies of data-processing adoptions in the late 1970s and 1980s indicate that both relatively low- and high-skill tasks were simplified or eliminated. Relatively low-skill keypunching and data-entry tasks were eliminated, for instance, when information could be entered directly into the computer via a terminal.

> A recent development still considered experimental but apparently quite successful, has been the use of 25 portable hand-held data entry devices by Finance Department water meter readers. Following the principle of capturing data as close as possible to its source, this new development has completely eliminated the former keypunch or data entry operator which consisted of transferring the water meter figures from paper to electronic medium.[32]

In addition, the evolution of machine languages — as well as the proliferation of software packages and user-friendly computer programs — has contributed to the increasing standardization and simplification of relatively high-skill programming tasks.[33] Such was the case in the early 1980s in a large insurance company as electronically based information became more widely accessible to decision makers throughout the firm.

> These people became increasingly able to access the information themselves. Not only were the data on-line, but the programmes needed to make use of them were now simple enough to employ, so that specialized

computer programmers were no longer needed as intermediaries between the user and the data.[34]

This case shows that the computer also took over some of the managers' supervisory and training tasks:

> Not only are there fewer people to manage for every routine aspect of insurance work, but the computer's built-in monitoring capability substitutes admirably for personnel supervision. In addition, computer-aided instruction is reducing the teaching work that supervisory staff used to do from basic instruction to handling merely the difficult cases.[35]

Moreover, the case studies indicate that as computers have become more sophisticated and as faulty components have become individually replaceable, many computer repair functions in offices (as in factories) have become easier. Complex repair problems are sent to the equipment manufacturer.

Lastly, the case studies demonstrate that skills created by a technology are vulnerable to deskilling as the technology matures. Relatively late adopters of data-processing technologies, for example, found that lesser-skilled computer operatives using prepackaged software programs could accomplish many tasks that previously required skilled programmers.[36] In addition, while early adopters of data processing cited strong demands for keypunchers and tabulators, later adopters did not. Case studies done in the 1980s also note decreasing demands for computer operators and relatively low-level computer data processors.[37]

CHANGING SKILLS, JOBS, AND OCCUPATIONS

The evolving nature of the demand for, and supply of, occupational skills as technologies mature influences the integration of changing tasks into jobs. The case study data highlight the effects of these labor market changes on existing jobs, on newly created positions, and on emerging occupations.

Job Enlargement

The case studies show that "job enlargement" occurs as firms pioneer adoptions of new technologies. Uncertainty in the quantity and quality of skill requirements, coupled with the unavailability of workers already trained in the newly emerging fields, encourages employers to add new tasks onto existing jobs—at least on a temporary basis.

The manufacturing case studies, for instance, highlight the tendency of early adopters of electronically controlled equipment to "enlarge" maintenance jobs to incorporate electronics maintenance and repair. The expanded job often required workers with a broad range of skills. With the introduction of automatic assembly in several factories in the 1950s, for instance,

> There [was] a need for a combination of maintenance skills in an "over-all machine repairman"—a man who can diagnose maintenance troubles arising out a combination of technologies. No firm claimed to have found a way of creating such skill. The maintenance executives simply searched hopefully—or desperately—among their maintenance men for this difficult combination.[38]

With the introduction of data processing into offices in the 1950s and early 1960s, some clerical positions were "enlarged." Although relatively routine tasks were eliminated, new computer-related tasks, such as data manipulation, report writing, and equipment operation, were added. A case study of a large U.S. electric power and light company gives an example of such enlargement:

> The work of the nonmechanized accounting groups responsible for the steps preparatory to the machine handling of the customer's account was completely reorganized. The specialized tasks previously done in five separate sections were consolidated into a "station arrangement," and each member of a station was trained to handle all five operations as part of a new, enlarged job. This reduction in job specialization resulted in increased efficiency in the allocation of manpower, and it is believed that most employees and their supervisors were more satisfied with these jobs.[39]

Another case study, of a large insurance company in the early 1950s, describes how clerical jobs were enlarged when computers were introduced into the commercial department:

After the conversion . . . a considerable upgrading in job content and skill of clerical jobs [took] place. . . .

The nature of the jobs of the clerks, in most cases, had completely changed. Prior to the conversion many clerks had been employed doing manual work of a detailed repetitive nature. For example, the clerks in one division had to sort and record manually 150,000 dividend notices per week. After the conversion, the detailed repetitive work of this function was done on machines, and the job of the clerks was either to control the accuracy of the work done on the machines or to operate the machines themselves.[40]

Early adopters of word-processing technologies enlarged their secretarial positions. In 1975, for example, word processing was new, so "there was no pool of operators already trained on the rather difficult word processing systems":

Word processing . . . has been seen as an additional skill to be added to those already possessed by clerical workers, rather than as a specialized full-time occupation. . . . As a result, wholesale restructuring of workflows and duties has not been necessary and word processing training has often been an occasion for the upgrading of secretarial work, rather than its deskilling.[41]

The Creation of New Positions

The earliest adopters of technologies are reluctant to create entirely new positions. When they find it necessary to do so — as is often the case when the adoption results in a substantial change in skill requirements — employers generally label the new positions "temporary," pending a trial period of several months.

As production techniques become relatively stable and skill demands more standardized, employers are more prone to add new types of jobs. The case studies show that early adopters of technologies generally incorrectly estimate the level of skills required for such new positions. More specifically, they indicate that, with few exceptions, firms overestimate the complexity of the new jobs, and the degree of skill and education actually required. One case study describes the problems resulting from the adoption of electronic data processing in a power and light firm in the late 1950s:

The new equipment . . . created two major classes of jobs: programmers and electronic equipment operators. Early in the change-over the glamor and complexity attributed to the new process affected employees' evaluations of these jobs. Both the jobs associated with the IBM 705 and the other newly created jobs were viewed as more complicated and thus open to higher job grades than the other jobs with which employees were familiar. Much of the subsequent dissatisfaction with the final job grades undoubtedly can be attributed to this exaggerated evaluation of the jobs. . . . Nearly everyone had at the outset overestimated the complexity of the new jobs because there had been no experience with them.[42]

The introduction of mechanized materials-handling operations in a pulp and paper mill in the mid-1950s also indicates overestimation of the degree of difficulty of new positions:

At first, management felt that the increased complexity of the new equipment, especially the new central control board, would require the foremen and controller to be high school graduates, and that employees in the remaining positions have at least an eighth grade education. After the woodroom had been in operation for a time, however, management reduced entry level educational requirements for the foremen and controller jobs, from a high school to an eighth grade education.[43]

Emerging Occupations

As uncertainty declines and demands for new skills become more widespread, new occupations emerge. The case studies portray the growing division of labor and changing demands for various occupations as a technology develops. In the manufacturing cases, engineers were expected to perform maintenance and repairs tasks as well as innovation and design functions during the early stages of development of a technology.

The technical sophistication of the new machines rendered necessary a further increased in the staff strength which, though small in numbers, sharply influenced costs. More engineers with a university or technical school training were needed than before for adjustments and repairs to the delicate mechanical and electronic control elements, as well as for supervising production.[44]

Adopters of similar technologies, years later, cited demands for technicians and maintenance workers to perform the tasks performed earlier by the engineers. For example:

The computer technician requires an extensive formal background and practical experience in electronics. He must be thoroughly acquainted with the capacity, capabilities and functional logic of the computer and with the electronic circuitry, electric and mechanical functioning of both computer and peripheral equipment.[45]

With more recent adoptions, operatives have been able to handle the maintenance and repair work. Repairs often involve replacing standardized components that were not available in the equipment of earlier adopters. With the change from batch to mass production, various technicians, operatives, and assembly workers supplant all-round machinists.

Unlike their early-adopting counterparts, late adopters of data processing and microelectronic technologies in the manufacturing industries hired workers into relatively specific computer-related occupations, such as computer programmers, computer technicians, and systems analysts.

The case studies show that as technologies develop and demand for new skills expand, experienced workers trained by other employers are available to firms undertaking technological adoptions. In addition, a supply of trained graduates from schools and colleges emerges and expands.

The range of training programs and credentials available from educational institutions also grows. Data-processing cases from the 1960s and 1970s, for example, cite the hiring of university graduates with computer training. More recent cases refer also to the hiring of graduates with computer skills from two-year colleges and vocational schools. Similarly, as the microprocessing technologies used in manufacturing matured, electronics skills were being taught not only in four-year colleges but also in community colleges and vocational schools.

JOB CHANGES AT THE WORKPLACE

While illustrating how general trends in skill needs evolve as technologies mature, the case studies highlight the need to address firm-specific characteristics in assessing the outcomes of technological adoptions on jobs at a particular workplace. The degree of change in the skills generated by the technological adoption, for instance, is cited above as playing an important role in determining whether current jobs are restructured or new positions added. More specifically, when relatively small changes in skill requirements are involved, established jobs tend to be redesigned. In contrast, a substantial change in skill requirements often precipitates new job classifications. The degree of skill change resulting from a particular adoption, however, depends on the firm's record of technological changes, skill and occupational mix, and prior allocation of tasks.

The case studies show that manufacturing firms handle new maintenance and technical skill requirements differently, depending on the prior degree of skill required. When relatively low-skill and unskilled hand processes were automated, new maintenance and technical job classifications were created. Prior to that time, only tools requiring little, if any, maintenance were used at these workplaces. In contrast, when highly skilled production work was automated, the maintenance and technical tasks associated with new electronic equipment were often incorporated into established jobs — jobs that previously had required relatively sophisticated mechanical and electrical skills for repairing large-scale equipment or precision tools.

A firm's prior skill and occupational mix also influences whether the shift in focus of worker tasks from setup and operation to monitoring and control leads to upskilling or deskilling. The cases show that when workers needed new "traits" such as attention to speed, agility, and alertness, the firms did not consider the changes sufficient to warrant the creation of new positions. When relatively low-skill operative functions were automated, such as in apparel and printed circuit board assembly, the operative jobs usually were enlarged and "upskilled." In contrast, when highly skilled operative work, such as that involved with continuous-process operations, was automated, the operative jobs were deskilled, as the new job attributes were viewed as less difficult than the types of skills previously needed.

Case studies of offices show that the skill level and organization of work prior to the change was also an important determinant in job changes. New clerical positions were established when word-processing equipment replaced traditional typewriters. In contrast, when firms adopting word-processing technologies had been using relatively sophisticated typing equipment, such as magnetic card IBM typewriters, existing secretarial positions were likely to be enlarged. Further, when typing had been done in offices by "typists," the adoption of word processing led to the creation of new "word processing operative" positions. But when typing had been just one component of a more diversified secretarial job, the new word-processing tasks became part of the existing position.

The manner in which new skill needs are introduced into the organizational structure also affects the impact of technological change at the workplace. The case studies indicate, for example, that centralizing new activities such as word processing or data processing in a separate department at the time of adoption — rather than dispersing them throughout the firm — promotes a greater degree of specialization and the likelihood that new positions will be created rather than current jobs enlarged. This decision seems to vary by firm size and depend, in large part, on the volume of demand within the firm for the particular activity.

EMPLOYMENT AT THE WORKPLACE

Technological change bolsters productivity and hence may contribute to additional employment in the firm. Alternatively, as technological change generally permits greater output with fewer workers, that is, "jobless growth," it may result in a contraction of employment.

Although helpful in portraying the processes and expected outcomes of technological change on skill requirements and particular jobs, the case studies do not help quantify the effects of technological change on employment level at the workplace. They do not isolate the specific impact of the technology on employment from that of other factors. Quite often the adoptions are timed to coincide with a business upswing in order to minimize the negative employment effects on the work force. The case studies' varying time spans as well as scopes of inquiry further limit the degree of aggregation and generalization that can be drawn about the impact of technological change on a firm's total employment.

Case study analyses of individual firms, however, do provide insight into several important factors. For one thing, the empirical evidence from the cases suggests that traditional, relatively aggregative measures — such as changes in average skill levels or in net change in employment at a firm — may do more to confuse than clarify the relationship between technological change and employment. Firms adopting similar technologies were found to experience different impacts on their average skill levels, for example, depending on factors such as the rate of growth, their prior occupational mix, and the timing of the adoption relative to the technology life cycle. The adoption of a newly developed technology may create more higher-skill than lower-skill jobs. Higher-skill jobs, however, often constitute a relatively small percentage of total employment at the firm. Hence expansion of the firm due to, say, gains in productivity could result in a *lower* average skill level overall.

The case study evidence also indicates that although the departments most directly affected often undergo a decline in employment as new equipment assumes labor-intensive tasks, overall net employment in the firms is affected in many different ways: some firms experience employment declines, some have net gains, and still others indicate no net change. Moreover, stability in terms of total employment generally masks considerable change at the workplace, as various jobs are created, eliminated, upskilled, or deskilled.

The case studies suggest that employers' decisions regarding the timing of technological adoptions played a key role in affecting net employment. If a firm introduced a technology during an expansionary phase of business, its net employment tended to increase. Frequently, when production facilities were expanded, new technologies were adopted; in some instances this was a deliberate choice, but in others the older equipment and systems were no longer available. Employment also increased in departments throughout the firm when technological changes were implemented to overcome a bottleneck in work flow.

In many of the cases, jobs did not expand as quickly as they would have had the technology not been adopted. The impact of this phenomenon — known as "silent firing" — on a firm's employment is particularly difficult to measure, as one can only speculate about what the level of output would have been without the new technology.

The case studies of firms in industries that were undergoing a long-term decline in output and demand — such as textiles, apparel, and coal mining — generally show employment declines after the technological change. Employers in these cases usually viewed the technology as a way to regain competitive advantage, and believed that even greater output and job losses would have occurred without that change.

Lastly, the case studies demonstrate the importance of *when* the impact of technological adoptions on employment is assessed. In several cases, productivity gains initially resulted in declines in the work force. Employment levels subsequently rose, however, as increases in productivity and efficiency resulted in expanded sales. In some instances, total employment remained below what it had been prior to the change; in other cases, the subsequent expansion overcompensated for the initial decline.

Moreover, employment statistics that are reported when firms are in transition can be misleading, as many such firms experience a temporary boom in labor demands. With many of the data-processing adoptions, for instance, firms required additional clerical workers to transfer large volumes of data from one system to the other. In addition, firms initially required a group of relatively skilled workers to analyze the systems and write the programs to automate other functions. After that, only program maintenance was needed to keep them running.[47]

SUMMARY

Providing empirical evidence across a wide range of economic and social circumstances, the case studies demonstrate that although the relationship between technological change and skill requirements is complex, it is neither random nor inconsistent. Firms adopting similar technologies, but at different points in the technology life cycle, face different skill and occupational needs. They also encounter dissimilar conditions with respect to the availability of workers with the appropriate skills.

The case studies permit the identification of common patterns and trends in the effects of technological change on skills, occupations, and jobs. These changing labor market conditions over the technology's life cycle affect the extent and character of the disruption posed by technological change at the workplace. A variety of firm-specific characteristics — including management practices, labor-management agreements, and firm size — influence how changes in skill requirements are incorporated into jobs at a particular worksite. The following chapter reviews the employment practices used in adjusting to the new mix of skills and jobs at the workplace, and examines how workers are affected by these adjustments.

APPENDIX 3A:
DISTRIBUTION OF CASE STUDIES

Table 3A–1   Distribution of Case Studies by Country.

Country	Number of Case Studies	Percentage
United States	108	54.8%
Great Britain	39	19.8
West Germany	14	7.1
France	7	3.6
Canada	7	3.6
Sweden	7	3.6
India	5	2.5
Australia	5	2.5
Other	5	2.5
Total	197	100.0%

Table 3A – 2   Distribution of Case Studies By Industry

Industry	Number of Case Studies	Percentage
Mining	5	2.5%
Construction	3	1.5
Manufacturing, total	117	59.4
Nondurable goods, total	46	23.4
Food and kindred products	12	
Textile and apparel	9	
Paper and paper products	7	
Chemicals	6	
Petroleum	8	
Leather	3	
Rubber	1	
Durable goods, total	71	36.0
Lumber and furniture	2	
Primary metals	16	
Fabricated metals	5	
Machinery, excluding electrical	6	
Electrical/electronic equipment	11	
Transportation equipment	21	
Instruments	2	
Stone, glass and clay	2	
Manufacturing (unidentified)	6	
Transportation, communications, and public utilities	21	10.7
Wholesale and retail	2	1.0
Finance, insurance, and real estate	29	14.7
Services	11	5.6
Government	9	4.6
Total	197	100.0%

4 STAFFING AND TRAINING PRACTICES OVER THE TECHNOLOGY LIFE CYCLE

In adjusting to technological change, employers rely on a variety of strategies, including retraining, the recruitment of new workers, transfers, liberalized retirement plans, hiring freezes, the use of temporary workers, and layoffs.[1] The technology life cycle, through its effects on skill requirements at the workplace and on the labor market for occupational skills more generally, critically influences these training and staffing practices.

Drawing on the experiences of the firms described in the case studies, this chapter takes an in-depth look at how employers integrate technological change with human resource adjustments at the workplace. It explores how these practices evolve over time, and what impacts these adjustments have on workers.

EMPLOYER ADJUSTMENTS TO TECHNOLOGICAL CHANGE

The case studies identify two major categories of human resource problems facing employers adjusting to technological change at the workplace: (1) meeting the new, relatively high-skill requirements generated by the change, and (2) reallocating displaced workers. The dimensions of these interrelated problems vary over the technology life cycle.

Meeting New Skill Requirements

Changing labor market conditions for skills as technologies evolve influence recruitment and selection practices of employers. Adopters of new technologies are often on their own in terms of assessing and meeting their new skill needs. By contrast, adopters of mature technologies face established occupations and credentials and supplies of appropriately trained workers to meet their new skill needs. Differences in recruitment and selection practices in turn affect the extent and content of training provided by adopters over the technology life cycle.

Recruitment and Selection Practices

When employers adopt newly emerging technologies, they are faced both with uncertainty about the quantity and quality of skills needed and with the lack of appropriately trained workers. As a result, employers often look to their current work force to meet these needs. The recruiting problems of early adopters of technologies are exemplified by the experience of a steel foundry that installed new continuous rolling mills in the 1960s:

> The building of the new plant gave rise to particular difficulties in the determination and satisfaction of manpower requirements. Some of the jobs had to be filled before they actually existed and before it was known exactly how production and working methods would work out. Even details of technical planning could not be final. During the initial period of recruitment of key personnel and the nucleus, there was still uncertainty about the extent of electronic equipment and automatic control to be expected.
>
> The works could only counter these uncertainties by using its existing staff as a recruiting reservoir for a flexible labour force of highly diverse skills, at least on a short-term basis.[2]

When existing positions are expanded to include the new, relatively high-skill tasks, the case studies indicate that in both factories and offices, workers occupying those jobs are generally upgraded. In manufacturing cases, for instance, maintenance and repair workers were upgraded when their jobs were enlarged to include electronics work. Relatively low-skilled production operatives also were upgraded as new control tasks were incorporated into their jobs.

Similarly, clerical workers were upgraded or promoted when their jobs were enlarged upon the adoption of data processing technology in the 1950s and early 1960s. The same was true of secretarial positions when word processing was introduced; in fact, several of the case studies on word processing suggest that the chances for upward mobility of those directly affected by the new technology was enhanced. In one such case:

> [The word processing system] has raised the professional status of the typist within the company, and also opened a different career path for her which could lead into office management or data processing.[3]

Another case indicates that the clerical staff who had been trained in word processing were subsequently promoted to positions in other departments. Such promotions, however, remained within the clerical ranks.

> Several lower level clerical staff have been promoted into higher positions (such as committee secretaries) not requiring active use of their word processing skills, while many of the others have moved onto similar work in other city departments where their word processing skills have been in high demand since the pace of training new operators has, until recently, fallen behind the increase in demand for these skills.[4]

According to the case studies, factories and offices differ considerably in how they staff new positions necessitated by technological change. Current employees continue to be assigned the new, relatively high-skill jobs in both types of workplaces; however, the pool from which these workers are drawn differs.

Factory workers who are displaced or whose jobs are deskilled generally are selected to fill the new, relatively high-skill positions. Labor-management agreements often play a role in this outcome. The case studies show, for instance, that while contractual agreements rarely contained specific clauses addressing how jobs created by technological change would be staffed, manufacturing workers generally were guaranteed job security if displaced from their regular positions. Because many of the manufacturing firms in the cases were experiencing sluggish employment growth, or none at all, there were few jobs into which displaced workers could be transferred. Moreover, although some modifications were allowed for skill needs, the fact that seniority was the major criterion used in allocating work assignments favored the displaced workers. This was especially true when seniority was determined within a department or production unit rather than throughout the plant or the firm.

When a new, computer-controlled eighty-inch steel mill replaced less complex forty-four- and seventy-six-inch mills in a large U.S. steel company, for example, the union contract gave employees who had been working on that production process first option in exercising their seniority rights to bid on the new jobs. This selection process was used even though the new equipment required considerably more sophisticated electronics skills than were needed in the former mills.

> More maintenance workers, such as skilled electrical repairmen and mechanical technicians, would be needed and existing skills would require substantial upgrading to enable maintenance workers to service the new equipment. Therefore, workers formerly employed as crane repairmen, millwrights, and pipefitters would have their skills upgraded through retraining for new jobs. Supervisory employees would also require additional training to enable them to understand the technology of computerized control and to adjust to the higher precision, productivity, and control requirements of the new facility.[5]

In contrast to the manufacturing cases, when new, relatively high-skill positions were created by office automation, the clerical workers most directly affected by the change rarely were assigned to the new jobs. Early adopters of computer technologies did not have the option of hiring skilled computer specialists from outside the firm. During the 1950s and 1960s, however, it was common practice to make lateral transfers of the clerical workers directly affected by the technological change, and to test-select all programmer trainees from *other* divisions within the firm. Interviews and references from management often supplemented these programmer-aptitude tests.

Moreover, it was not unusual for the job descriptions of relatively low-level clerical workers to be rewritten when they left, and for the new position to be "filled with more highly skilled professional and technically-oriented people."[6]

The long-term pattern is quite clear. As lower-level clerical positions become vacant through attrition, these positions have tended to be upgraded to job classifications requiring more sophisticated analysis and interpretation, rather than just a mechanical handling of information.[7]

Undefined job ladders and promotion patterns complicated the staffing of new types of positions created by office automation. In offices adopting data processing prior to 1960, for example, some employees were reluctant to move into newly created departments for fear of being removed from the mainstream operations of the firm and "known" career ladders. The adoption of an electronic data-processing system in a large oil company in the 1940s highlights this internal mobility problem prior to the development of computer-related career paths:

One of the problems with which we have been faced is that of selecting and training computer personnel. Several aspects of this problem revolve around the lack of clarity concerning computer work as a career. We have not yet established career lines which show where a person goes after a successful period in computer work. As a result, some people, when asked to enter the field, have declined on the grounds that it would remove them from the "mainstream" of the company. Others have said they do not want to be branded as "machine men" when their careers are in economics, accounting, plant administration, personnel administration, technical or scientific fields, etc. Yet these are often the very people we want, because of their subject matter experience and other qualifications.[8]

The emergence of new occupations, which occurs as demands for various skills increase, appears initially to be a mixed blessing for employers seeking to meet new high-skill needs. Firms adopting technologies at this time are considerably more certain than earlier adopters about the quality and quantity of skills and workers they need.[9] Moreover, these employers are able to hire skilled workers from schools and colleges. In addition to reducing the firm's training costs, hiring college-trained graduates minimizes the drain of key personnel from other divisions of the company when staffing new positions. When new occupations emerge, however, adopting firms often face shortages of skilled workers. As growing demands outpace the supply of appropriately trained workers from educational institutions, firms aggressively "pirate" experienced workers from other firms to meet their new skill needs.

Lastly, employers adopting relatively mature technologies often recruit and select workers to perform the new skill needs from outside the firm. Job enlargement and skill shortages were rarely mentioned in these case studies on late adopters.

Training Practices

The empirical evidence from the case studies supports the notion of a skill training life cycle in occupational preparation and illustrates the changing extent and nature of skills provided by the employer as a technology matures.

Training is one of the most significant problems facing employers involved with the adoption of technologies that have not yet matured. As indicated in the case study of a light and power company adopting electronic data processing in the early 1950s:

> Throughout the change period, supervisors complained both about the lack of time for adequate training and about the large amount of time spent in training. Not only were there many replacements on old jobs who had to be trained, but old employees assigned to new tasks also had to be retrained, frequently on overtime. Training for new jobs was particularly difficult. For example, the first employees on new jobs in programming often had to be trained on a process using materials which were unfamiliar to both the supervisor and the trainee.[10]

When current employees in factories and in offices are selected to staff the higher-skill jobs created by technological changes, they usually are trained in employer-sponsored programs. Firms operating at the "cutting edge" of new technologies, as did this aircraft manufacturer, have little alternative:

> [The company] does not teach skills when applicants with these skills can be hired. However, intensive in-plant training has been a necessary companion of the great forward strides made by our engineers. For the past ten years the company has been working in fields never before explored by industry. It has been impossible, in most cases, to hire people with the required experience. Teaching the skill has been the only answer. . . . More than 90 per cent of [the company's] requirements for skilled technicians have been met by hiring relatively inexperienced persons and training them upward through a succession of skills.[11]

According to the manufacturing case studies, early adopters of computerized production and instrumentation devices usually provided workers with formal classroom instruction in electronics and computer control systems. Informal on-the-job training followed. In the aforementioned steel company that introduced an eighty-inch steel mill, for example, "a training program for about 400 hourly workers and supervisors was initiated between one and two years before the mill was completed."[12]

Firms that failed to prepare their workers for the newly required electronics skills did so at the risk of excessive, very expensive "downtime." After a few highly disruptive breakdowns in newly installed, sophisticated equipment, for instance, a bakery had to produce a "crash" training program course for its maintenance staff, for whom "there was not time available" earlier to train.

> The weight of responsibility fell on the shoulders of maintenance men who, in contrast to the process operators, had been poorly prepared for their work. The electronic controls, control circuits and microswitch arrangements were new to the electricians, and new forms of variable speed drives, finer tolerances and new materials were unknown to the engineers, so that reliance had to be placed on the services of suppliers. The trained and, this time, skilled operators were ineffective against blown valves or fused circuits, or broken timing belts or internal parts. . . .

> A "crash" training scheme designed to provide an overall picture of the production process in theory and practice was necessary for the maintenance section. This course involved the intimate study of circuit drawings and engineering plans, an actual practical examination of all major parts involved and a thorough breakdown of most components.[13]

The office automation cases also indicate that firms tend to retrain their own workers when jobs are restructured or new jobs are established during the relatively early development of a technology. They often rely on equipment manufacturers to train employees in the new skills. Such training programs are often extensive. For adopters of automated data processing in the 1950s, for example, training often began months prior to arrival of the equipment:

> Training for the programming and operating staff consisted of sessions at the manufacturer's schools for several weeks, attendance at electronics seminars, visits to other companies, and a great deal of study. The "curriculum" included such technical systems work as the designing of flow charts, the construction of forms for both input and output, the intricacies

of punched cards, the elements of the particular machine language and the logical processes of the computer. The balance of the training was done on-the-job in actual programming, analysis of applications, debugging, and helping with the installation of the computer.[14]

Similarly, in an insurance firm during that period, current employees received considerable training from equipment manufacturers.

> The formal training course preparatory to large-scale computer operating was an eight-week course conducted by the computer manufacturer in New York. The classroom instruction covered theory and computer logic, programming, operating instructions and binary arithmetic. The practical part of the course consisted of practice operating on a large-scale computer on the premises of one of the customers of the computer manufacturer. This practical work took place three nights a week from midnight to 7 a.m. from a period of five weeks.[15]

Due to the limited demand within a given plant for computer maintenance and repair functions, and the relatively high level of skills required, most early adopters of data-processing technologies contracted out these services and training needs to the equipment manufacturers. When these firms — especially if they were large — did hire full-time computer technicians, it was the equipment manufacturers again that provided the training. In a large insurance company that introduced electronic data processing (EDP) in the 1950s,

> these new assignments [computer technicians] were preceded by extended periods of formal classroom and practical training that exceeded in duration and depth any training taken by any of the other EDP occupational groups, including the programmers. The course in Computer Logic and Maintenance lasted five months and was taught by computer manufacturer's instructors on the computer manufacturer's premises. This course did not include instruction in basic electronics, a thorough knowledge of which was assumed on the part of the candidates. The course content covered the logic of the functioning of the computer, the electronic circuitry of the computer, and practical techniques used in preventive maintenance and trouble shooting. Later, the technicians took an additional four-week course on the maintenance of peripheral equipment components — the card-to-tape converter and the high-speed printer.[16]

With the early adoption of word-processing technologies, employees also were retrained by the manufacturer of the equipment. This formal training was followed by on-the-job learning.

Current executives and managers that were selected for the new, high-level data-processing jobs also were taught computer-related skills in company-sponsored programs. As noted earlier, this practice of recruiting individuals for newly created positions from other divisions of the firm drained key staff from other company functions. Nevertheless, firms generally found it more effective — at least during the initial conversion — to teach computer technology to their high-ranking managers and executives than to attempt to teach their business — be it insurance, banking, airlines, or whatever — to computer specialists.

> The top EDP jobs — administrators, project planners and systems analysts — are generally filled from within the organization itself. Those selected for these positions are usually chosen for their subject matter knowledge and experience in methods and systems work. Electronic data processing expertise is usually acquired subsequently. The manufacturers of computing equipment often provide expert assistance in this work. Occasionally, the services of a management consultant will be retained.[17]

The case studies indicate that firms often find it more efficient to provide an increasing share of computer-related and word-processing skills themselves over time, rather than to depend on the equipment manufacturer. This practice permits employers to streamline training to match their needs more exactly, and to place more emphasis on the requirements of their work by "slanting the training towards those features of the machine that are of most importance in . . . operations."[18] Employers also gain greater flexibility, as they can schedule the training during relatively slow periods in the worker's day. Employees trained earlier by the manufacturer frequently help set up and staff these company-based training programs.

The availability of appropriately trained individuals in the external labor market reduces the probability that the firms undertaking technological adoptions will train their workers. As noted earlier, as skills become transferable and firms begin hiring experienced workers from other firms, the expected return to such training investments falls. As demands for particular skills expand further and the supply of appropriately trained graduates from schools and colleges increases, firms prefer not to provide this training. In addition, high turnover rates of workers in new and emerging occupations encourage employers to narrow the scope of training to meet their firm-specific needs.

The case studies show that as demands for electronic skills increased and colleges and vocational schools expanded their curricular offerings in this field, firms adopting electronically controlled manufacturing processes expected new hires to ac-quire their basic electronics in college courses. Whereas early adopters had taught these high-level skills—there being no other source of supply—in the early 1980s employers in need of workers skilled in electronic technology emphasized the impor-tance of formal educational in this field. The experience of a large public utility firm was typical:

> "In the old days," . . . [the director of engineering] says, "engineers often came up through the craft ranks. Now we're looking for people with at least a bachelor's degree in electrical engineering, math or science. We still promote craft people, but they have to take a lot of continuing education to catch up."
>
> The reason engineering jobs are requiring ever-higher levels of formal education has to do with the nature of electronic technology. . . ." When state-of-the-art equipment was all either mechanical or electro-mechanical, working with it was an effective way to learn it," [the director of engineer-ing] says. "But computer and electronic technology is harder to grasp—formal education is essential to understanding it."[19]

This firm did continue to run courses in-house, for its engi-neers, on "state-of-the-art technological advances."

The case studies also highlight the shift in the provision of computer-related skills from the workplace to the schools. A case study of a large, technologically sophisticated insurance company, for instance, notes the change in its source of pro-grammers over time:

> The normal source of persons for programmer positions now is from outside the subject company. University graduates are being recruited. This is a change in company policy. Early in the conversion period to electronic data processing, the usual company practice was to select currently em-ployed personnel for training programs.[20]

By the 1970s, firms installing data processing were recruiting college-trained graduates for a range of computer-related jobs, including computer programmers, analysts, technicians, and the like.

Reallocating Displaced Workers

Throughout the technology life cycle, workers are subject to job displacement — posing the second major set of human resource problems facing employers adopting technological changes. With very few exceptions, workers were displaced by each of the technological changes discussed in the case studies. As indicated previously, when data-processing technologies were introduced, the workers displaced rarely were promoted to the better jobs created. Early adopters of manufacturing and word-processing technologies, often promoted the workers whose jobs had been simplified or eliminated — however, the number of better positions created generally fell short of the number of these workers. Moreover, displaced workers in late-adopting firms often lacked formal educational credentials needed for the newer, skilled jobs.

Most firms cited in the case studies took steps to reduce the pool of workers seeking reemployment within the firm. For instance, firms generally sought to adopt technological changes during periods of business expansion and economic prosperity. Growing demand within the firm increased the likelihood of job creation. In addition, high labor demand in the area increased voluntary quits and helped minimize the amount of adjustment to the adoption required within the firm. When employers encountered relatively low rates of attrition among employees — which was the case when high-skill manufacturing work was automated — they often established liberalized retirement plans and severance pay packages to discourage workers from seeking reemployment in the firm. In addition, several firms implemented hiring freezes or used temporary workers in the period prior to the adoption to minimize the amount of disruption among regular employees.

When unable to reassign all workers seeking to remain with the firm, employers often offered to help them find alternative employment. Some of the more unusual strategies identified in the case studies include starting a subsidiary firm to absorb these workers;[21] setting up a trust fund to subsidize the entry of new firms into the community;[22] keeping redundant, unproductive workers on the payroll until retirement;[23] and creating a "reserve pool" or "shadow posts" for displaced workers until an appropriate job became available.[24]

Lateral Transfers

The case studies suggest that most workers who are not promoted after being displaced, but who want to remain with their employer, are absorbed by growth or replacement needs elsewhere in the firm. Most of these workers are laterally transferred and require little, if any, additional training. Workers whose jobs were eliminated with the automation of relatively low-skill manufacturing tasks generally were moved to similarly classified jobs in the firm. Lateral transfers also were common for displaced clerical workers. These clerical workers were transferred either to other existing office jobs at the firm, or to the relatively low-skill jobs, such as keypunching and tabulation, created by data-processing adoptions.

With the adoption of electronic data processing in a large insurance company in the early 1960s, for example:

> Almost all training/retraining of EDP-affected clerical workers was of the "understudy" or on-the-job type. . . . The major exception was and is the released-time typing refresher course that improves present employability and makes future reassignment easier.[25]

Similarly, with computerization of a large bank in the early 1960s, bookkeeping machine operators became keypunch operators. Other clerical staff whose positions were eliminated were transferred into the additional positions of junior accounting clerk, return items clerk, and floater clerk, and the newly created position of data examination clerk and data examination clerk supervisors.[26]

Downgrading

Instances of downgrading cited in the manufacturing case studies resulted primarily from the automation of relatively high-skill production work. The downgraded workers were primarily machine operatives who were passed over for the new, higher-skill tasks created, but who remained in their deskilled jobs due to a lack of alternative employment opportunities at the firm.

Downgrading also occurred with office automation, particularly as a range of computer-related occupations emerged. This was the case, for instance, when the tasks of clerical workers became deskilled and the computer or workers in new job classifications absorbed some of their accounting or bookkeeping tasks. The remaining jobs became more routine, involving a smaller range of easier tasks, such as answering the phone, typing, and photocopying. Downgrading also occurred when some supervisors and lower-level office managers were demoted after their positions were deleted and they were bypassed for the new, higher-skill jobs.

Generally, workers retained their former wages when downgraded. Some firms, however, particularly those in manufacturing, viewed such a practice as disruptive to overall job and wage structures, and hence reduced the workers' wages. In addition, several of the firms that adopted office technologies "red-circled" (froze) the wages of the downgraded workers. With future pay increases restricted, employees are likely to experience a decline in purchasing power over time. This was the case, for instance, in a Canadian city government agency that was reorganized following adoption of electronic data processing:

> The salary of employees in such a situation is frozen (red circled) at the nominal level received before reorganization. The buying power of that salary is thus allowed to erode at the current rate of inflation, effectively reducing the real wages received.[27]

Layoffs

The case studies suggest that relatively few workers were involuntarily displaced from the firm when technological changes took place. One-quarter of the case studies indicate that layoffs occurred after the adoption of the technology. In most of these cases, less than 1 percent of the firm's total work force was laid off.

In a few cases, however, layoffs were extensive. In particular, substantial layoffs were associated with technological changes that took place in "declining industries" or that involved the relocation of a plant to another geographic area. For firms in industries experiencing a long-term decline in output and demand—such as textiles, apparel, and coal mining—transfer opportunities were few, if any existed at all. Hence most of the workers whose jobs were eliminated by technological changes in these industries were laid off.

In addition, several plants in the case studies reported considerable layoffs after they relocated due to the opening of a new plant or an intrafirm consolidation. The case studies indicate that workers often refuse to transfer if a long commute or a residential move is necessary. Thus the level of activity in the local economy plays a critical role in influencing the extent of layoffs when a plant closes.

Plant closings or considerable employment cutbacks due to technological change in "company towns" are particularly likely to generate widespread layoffs. The case studies show that even firms that exhibit highly paternalistic relationships with their employees—which is common in firms that dominate a local labor market—could not compensate for the lack of alternative employment opportunities in the area.

A case study on the introduction of the "float process," whereby workers were virtually replaced by machines in the production of flat glass, demonstrates the vulnerability of an area whose employment base is tied to one production process.[28] At the time of the adoption of the float process, the multinational Pilkington Brothers glass-manufacturing firm, the largest of three glass manufacturers in St. Helens, England, accounted for 20 percent of the total work force in the area in the 1970s. Many of the employees were the fifth and sixth generation of their families to work for Pilkington. Beyond its role as provider of jobs and wages, Pilkington—"the godfather of St. Helens"—greatly affected the private lives of its past and current employees. The firm provided a wide range of recreational facilities, including bowling greens, tennis courts, cricket, rugby and football pitches, and snooker and dart facilities. Moreover, for pensioners the company paid for part or all of "the cost of new spectacles, false teeth, television licenses, heating and even seeds for their gardens."[29] Meals and visits, at hospital or home, were also provided for invalid pensioners.

The effects on the community of employment cutbacks at Pilkington were considerable, in spite of some valiant attempts by the company to create new jobs in the area for its workers. Pilkington, for instance, had invested in a new subsidiary whose purpose was "to provide venture capital and expertise to would-be entrepreneurs prepared to launch new business in St. Helens."[30] The company's management also supported a trust that provided interest-free loans and free technical assistance to local firms.

DISTRIBUTIONAL EFFECTS ON WORKERS

Technological change at the workplace has disparate effects on individual workers. Some benefit by being hired for new jobs, or through upgrading and promotion opportunities created by the changes. Other workers are hurt through the deskilling of tasks, the elimination of jobs, and the disruption of established career ladders.

Empirical evidence from the case studies show the uneven distribution of these outcomes. Workers at all skill levels were found vulnerable to the negative impacts of technological change; however, those who were laid off or downgraded had the least potential for obtaining good, alternative employment.[31] When lesser-skilled and unskilled work was eliminated, employees who returned to the labor market were qualified for only the lowest-skill and lowest-paying jobs. Similarly, highly skilled employees laid off or demoted, particularly in declining manufacturing industries, found their extensive work experience and skills of little value in seeking other jobs.

Gender Effects

The case studies suggest that female workers benefit far less from technological change than do male workers.[32] In general, the cases portray a view that there are "male jobs" and "female jobs," identified along very traditional lines. These patterns are reflected in many of the employers' recruitment and staffing practices. In a firm looking for "suitable applicants" to fill new word-processing positions, for instance, one manager noted:

> We looked predominantly for young women, perhaps between 18 and 25 years old, who had a specific aptitude for typing and typing equipment and who were team girls whose personality would fit easily into an interdependent daily routine.[33]

When technological change eliminated some jobs and workers had to be transferred or laid off, women often were assumed to be less "appropriate" for or less in need of the remaining positions. For example, in a firm that had a policy of providing employment security to workers displaced by technological change, an "employment reserve" was established in which displaced workers were given temporary assignments until a permanent position became available. Overall, the reserve worked quite well: new jobs eventually were found "for almost all employees displaced by technological change."[34] Displaced females however, ran into problems:

> The employment reserve was unable to absorb approximately 30 female inspectors, who were laid off from inspection jobs in the tin mill. . . . There are very few production jobs at [the company] for women and this was the prime reason for the layoff. The problem was sex and not the employment reserve.[35]

Another case indicates that although there was no strenuous or difficult activity involved in the newly created positions, women were not considered, for plenty of males who were "primary breadwinners" needed the work.[36]

When the adoption of a technology resulted in additional shift work, male employees generally were preferred for the new night shifts. Males were given priority for a variety of reasons, including a higher absenteeism rate among women workers, especially at night; laws restricting the use of women at night; and transportation problems particularly for women on nights and weekends.

In one case, however, where many women were viewed by management as the "primary breadwinners for their families," they were given a trial on the new, seven-day-a-week, twenty-four-hour-a-day, three-shift system. The understanding was that "if it didn't work out we just call the whole thing off and hire males." The women performed well on all shifts, causing management to exclaim that "they surprised us, they surprised the union, they surprised themselves."[37]

Some plant managers suggested that the effects of automation were better for women than for men, as it simplified much of their work. This was believed to be the case, for example, when automation deskilled operating jobs in an auto parts firm:

> The motivating force for women is job conditions. A properly mechanized job to them means a cleaner, nicer job with more prestige. *More automation makes it easier to do the work* [italics in original]. That means it's easier to keep help. With men it's different. We want men to progress to more skilled jobs to become technicians, job setters, and potential supervisors. This means they need to have initiative. I'm not so sure that automation encourages that initiative.[38]

Moreover, the "traditionally male" jobs that were "given" to women after a technology was adopted were the deskilled jobs that males refused to fill. For example, with the introduction of the automatic buffing machine in a bumper-plating firm:

> Direction of the automatic buffing machine and portions of the plating cycle were turned over to women operators, since they were required only to man simple control boards or inspection stations. Automation absorbed the effort and skill in buffing and the effort of inspections.[39]

Although "emerging" occupations had no gender stereotype before they were filled, the better, higher-skill positions were almost exclusively staffed by males.[40] In contrast, females staffed the new, lower-skill jobs. This pattern applied even when women constituted all the workers displaced by the technological adoption.

For instance, in several firms in the insurance industry—an industry often at the forefront in adoptions of office technologies—almost all the new computer systems analysts and computer programmers were male, while computer operatives usually were female.[41] Men were found to have the "necessary aptitude" for the better jobs, whereas, employers noted, women make very successful computer operators.

> This is not to say that female staff cannot cope with this work [systems and programming] but the necessary aptitude seems to be found in a much higher proportion among males than among females. . . . We have, however, employed female operators from the outset with marked success.[42]

Similar results are seen in the public sector. When a regional office of the Internal Revenue Service introduced data processing, for example, the vast majority of the workers employed in the units directly affected by the change were female, but 95 percent of the workers chosen for the new data-processing positions were young men from other departments.

> Key ADP jobs were filled through a systematic selection procedure which included written tests, interviews and supervisory evaluations. At first, only a few employees from district offices whose jobs were scheduled for elimination applied or were selected for ADP jobs, apparently because they lacked the necessary motivation to relocate and retrain.[43]

In most of the data-processing cases, when the new, relatively high-skill positions were filled internally, general aptitude tests, supplemented by references from and interviews by management, were the bases for selection of these personnel. It is not possible from the case studies to determine the relative importance of these factors, or even if females applied for these jobs. In some cases, managerial references and interviews were the only methods used in the selection process. In addition, the practice referred to above, of firms teaching computer-related skills to their top managers and executives, effectually excluded females from acquiring the new, higher-skill data-processing positions. Employees in the top managerial and executive positions in these firms were almost exclusively males.

The net result of these selection practices is a lack of women in the new, skilled positions created by technological change. Men were generally determined to be "better suited" to fill the better jobs, and they were subsequently provided the new skills required of the jobs in company-sponsored training programs.

Age Effects

Older workers encounter far more difficulties than younger workers in adjusting to technological change at the workplace.[44] Studies of firms that automated highly skilled manufacturing work frequently cite the age of the workers as a key problem in the adjustment process. These cases, which involve a relatively older work force—a characteristic of many of the traditional manufacturing industries—indicate that the older workers were often unable or unwilling to be retrained for the new jobs. Promotions to the new, higher-skill jobs required a significant amount of retraining in an area very different from the ones in which these workers had already invested many years in apprenticeship programs and on-the-job training. In addition, in many of the manufacturing case studies, new operative tasks necessitated increased speed and attention to detail. Older workers often were hard-pressed to meet these new requirements.

In several cases, management expressed specific preference for younger workers to fill the new, higher skill jobs. Older workers, they said, generally required more training than their younger counterparts or were too rigid. In the selection of employees to staff the better jobs following the introduction of electronic equipment in a bakery, for example, experience had to be coupled with "sufficient youthfulness (preferably under 40 years of age) for quick absorption of theories and techniques involved in the process."[45] Similarly, in a large oil company in the 1940s,

> The available resources from within the company are not unlimited, and considerable care must be taken lest we employ exactly the wrong kind of people for systems or computer work. By the "wrong kind of people" I mean those who, because of basic rigidities or years of exposure to old ways of doing things, cannot apply imaginative thinking to the development of data-processing systems.[46]

A case involving the introduction of an electronically controlled materials-handling system in a British factory raises the issue of "hierarchical estrangement" for older workers:

The influx of computer technology and its associated sophisticated electronics systems and techniques have resulted in a rise in the number of graduate managers and specialists in OR, systems analysis, and so on. While this has served to permanently introduce these new techniques into the company in such a manner that their practitioners are now having considerable influence in decision-making and planning, it has also tended to highlight the problem of "hierarchical estrangement" for the older, "shop-floor" managers and supervisors who have to increasingly deal with complex technology and with younger supervisors and who, perhaps, have less inclination to accept new management techniques.[47]

In other instances, older workers were viewed as simply "non-retrainable." In addition, the case studies suggest that when technological change is coupled with relocation of the worksite or with additional shift work, older workers generally are more reluctant than were their younger counterparts to make the necessary adjustments.

Age was rarely noted as a problem in the case studies on office automation. The work force affected by the technological changes tended to be considerably younger than that in the manufacturing cases. Moreover, the relatively high rate of turnover of the young clerical workers facilitated the firms' adjustment processes by providing a continuous array of job openings. When older workers were affected by technological change in offices, they usually were reemployed in other departments. Moreover, in a few of the data-processing cases, the firms indicated that they retained older workers until their retirement even though no appropriate job was found for them.[48]

Blue-Collar and White-Collar Effects

Blue-collar and white-collar workers also are affected differently by technological change. For one thing, the case studies suggest that blue-collar manufacturing workers are more likely to be laid off than white-collar office workers when firms are automated. The lack of layoffs in the case studies on office automation is attributable to three factors. First, office automation generally occurred in firms in the relatively rapidly growing, nonmanufacturing sectors of the economy. Thus reemployment within the firm was facilitated by job openings due to growth. Second, those firms rarely relocated their worksite, so the need for workers to move or commute long distances was minimized. Third, the firms were characterized by relatively high rates of employee turnover. This high turnover rate diminished the size of the pool of workers seeking reassignment; it also generated a considerable number of jobs into which dislocated workers could be transferred.

However, while less prone than factory workers to be laid off due to technological change, displaced white-collar office workers were less likely than their blue-collar counterparts to be promoted or upgraded to the new, high-skill positions. Different "aptitudes" generally were required before and after technological changes in both factories and offices. However, while employment growth made lateral transfer highly feasible in offices, sluggish employment, coupled with labor-management agreements providing job security and reassignments based primarily on seniority, fostered upgrading in manufacturing.

Growing Disparity Between Gainers and Losers

The case studies suggest a growing disparity between gainers and losers from change as technologies mature. As shown earlier, current employees are often selected and trained to perform new, high-skill tasks generated by the adoption of new technologies. In contrast, as technologies mature, the emergence of new educational credentials and occupations, combined with a growing supply of appropriately skilled workers, encourage the hiring of trained specialists from outside the firm. Thus changes in staffing and training practices over the technology life cycle foster barriers to upgrading and discontinuous job ladders for workers.

A study in the early 1980s of a large transportation and communications firm notes that only 2 of the 130 displaced clerical workers moved up to the professional or managerial ranks due to a "growing skills, educational and aptitude disparity between the two occupational levels."[49]

The increasing use of academic credentials to fill the high-skill positions generated by the introduction of computers in public-service employment in the 1970s and 1980s also diminished the

opportunities for promotion of many lower-level clerical workers, as a case on the Toronto city government shows:

> The work of lower-level clerks, when not eliminated entirely, has been
> radically altered — some would say deskilled — by these systems, while the
> work remaining for higher-level clerks now requires more advanced and
> analytical and accounting skills than ever before.

> An examination of changes in the educational requirements of Finance
> Department employees from 1970–1980 reveals a marked increase in the
> amount of formal education required of higher-level clerical workers and
> their immediate supervisors, as well as profound changes in the way
> routine tasks are handled by lower-level clerks.

> In 1970, the only departmental educational requirement was completion of
> secondary school equivalent attainment. By 1980, many clerical positions
> required considerable advanced training at the post-secondary level as well.[50]

Similarly, the case of an insurance firm that "changed from a
pillar of clerical employment to a pillar of professional and
specialist employment" during the 1970s indicates a "widening
skills gap" occurring between the low-level clerical workers and
the newly hired professional and technical employees.

> There appears to be a widening skills gap between what is considered
> clerical and what is considered professional work. Of the job openings
> posted internally during the first six months of 1980, only half resulted in
> internal candidates being chose. The reason for this low internal promotion
> rate was, according to a company personnel official, "the growing number
> of specialist positions demanding previous related experience."[51]

SUMMARY

Empirical evidence from case studies shows that human resource adjustments to technological change at the workplace
focus on meeting relatively high-skill needs and on reallocating
workers who are displaced. Employers' recruitment and selection practices to meet their new skill requirements are found to
vary systematically over the technology life cycle. A skill-training life cycle evolves, highlighting differences in training needs
and in the institutional providers of job skills as techniques
mature.

Employers use a variety of strategies, including promotions, early retirement plans, lateral transfers, downgrading, and lay-offs, in response to worker displacement. The case studies highlight the importance of an expanding business or local economy in easing the firms' adjustment process by providing alternative employment opportunities. They also show, however, that a booming economy does not ensure against technology-induced layoffs and reemployment difficulties. Technological change in declining industries and those involving work site relocation, in particular, generated extensive adjustment problems, regardless of the state of the economy.

In the aggregate, the studies suggest that most workers displaced by technological change are promoted or laterally transferred within the firm. Analysis of the individual case studies, however, highlights the difficulties of the less fortunate workers who experience downgrading, layoffs, or involuntary relocation. The cases demonstrate the unevenness in the distribution of benefits and costs of technological change, with displaced women and older workers faring less well than their male or younger counterparts. The case study evidence also suggests an increasing disparity over the technology life cycle between those who gain and those who lose from technological change.

III TECHNOLOGICAL CHANGE AND THE COMMUNITY

5 TECHNOLOGICAL CHANGE AND THE LOCAL LABOR MARKET

The production life-cycle framework and the empirical evidence in the preceding chapters suggest how technological change affects a firm's labor, skill, and training requirements. The mix of firms and their production activities in an area will in turn influence the impacts of technological change on the labor market in a particular local economy.

This chapter looks at the human resource implications of these processes through a detailed examination of a technology-based reindustrialization of a local economy. The focus is on the labor market in Lowell, Massachusetts, an economically depressed community that has been transformed into a thriving center of high-technology employment. Located twenty-five miles northwest of Boston, the Lowell area is often viewed as a "model of reindustrialization" for older cities and regions throughout the industrialized world that have lost jobs in their traditional manufacturing industries.[1] (See Figure 5–1.)

The chapter begins with an overview of the Lowell economic turnaround, highlighting the circumstances surrounding this transformation and the key roles played by local labor market conditions and human resources. It then reviews how the mix of employment changed by industry and occupation as Lowell's employment base shifted. The chapter concludes with an analysis of the human resource experiences of twenty-six firms in the area.

Figure 5-1 The Lowell Labor Market Area.

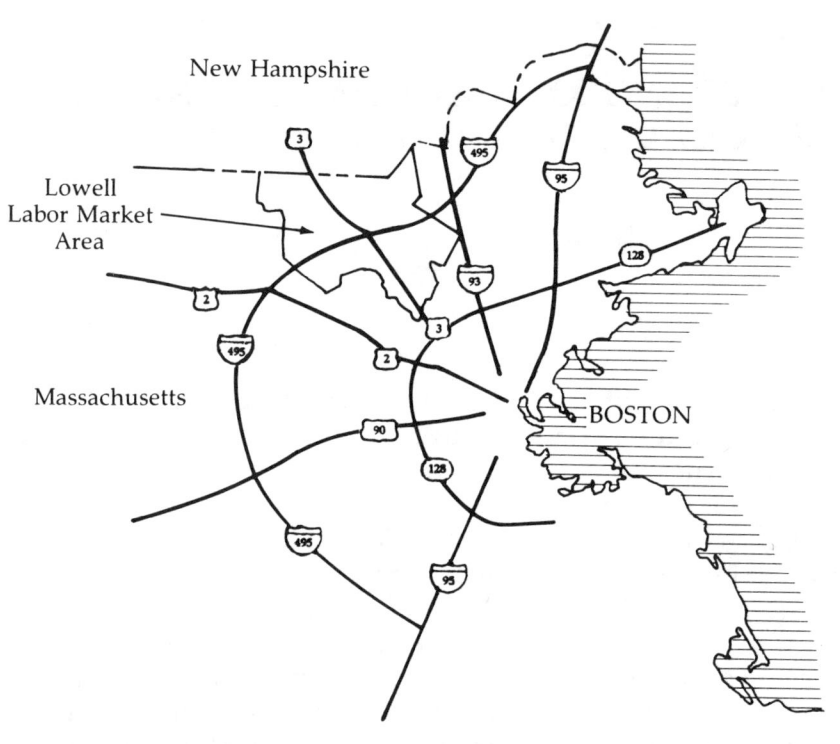

THE TRANSFORMATION OF THE LOWELL ECONOMY[2]

A Brief History

An infusion of textile mills in the nineteenth century transformed Lowell, Massachusetts, from a sparsely populated rural area into the state's second largest city and the heart of the U.S. textile industry.[3] From 1826 to 1850 Lowell's population expanded from 2,500 to over 33,000; by 1920 the population had more than trebled, to over 112,000.

The impetus behind this growth was manufacturing—specifically textiles. Employment data for the textile industry in Lowell in the nineteenth century were not available; however, production statistics for the 1835–1888 period indicate a substantial growth in the number of jobs. For instance, the number of textile mills in Lowell rose from 22 to 175, and the output of cotton cloth rose from approximately 750,000 to almost 5 million yards per week. By the early 1920s, employment in the textile industry accounted for over 40 percent of all manufacturing employment in Lowell.[4]

Lowell's dramatic growth came to an abrupt halt in the late 1920s. Manufacturing employment in the city fell almost 50 percent between 1924 and 1932. This decline was partially the result of the Great Depression. In addition, however, much of the textile industry in the New England region shifted to the South during this period. Hence when the national economy recovered from the depression, Lowell did not. Instead, the local economy stagnated for the next three decades.

In the 1960s the Lowell economy witnessed a revival that spilled over into the entire labor market area. Throughout the 1960s and 1970s, the local population and employment expanded more rapidly than that of the state and the nation. By 1980, the Lowell Labor Market Area—which includes the towns of Billerica, Chelmsford, Dracut, Dunstable, Tewksbury, Tyngsborough, and Westford, in addition to the city of Lowell—had a population of 227,000. Local employment growth was particularly rapid after the nationwide recession in the mid-1970s— growing over 6 percent a year on average from 1976 to 1982, a rate more than double that of both Massachusetts and the country overall.[5]

This relatively fast employment growth significantly reduced the local unemployment rate compared to that of the nation. The historically large gap between Lowell's unemployment rate and the national rate began to narrow in the late 1960s; it closed considerably after the recession in the mid-1970s. Since 1979 the Lowell area's unemployment rate has been below that of the United States. In 1986, the unemployment rate in the local area was 4.0 percent, compared to 7.0 percent nationwide.

Employment by Industry

Although the Lowell area maintained its tradition of being more oriented toward manufacturing than either Massachusetts or the nation, the manufacturing base in the local economy declined in both absolute and relative terms during the 1960s and early 1970s. In the early 1970s, however, employment patterns within Lowell's manufacturing sector became mixed. Significant employment losses continued in the nondurable goods manufacturing sector, but several durable goods industries experienced rapid growth. The electrical and electronic equipment industry, for example, grew at a rate of over 9 percent a year between 1970 and 1976, a rate more than three times that of total employment in the Lowell area. The nonelectrical machinery industry also experienced above-average growth.

In the late 1970s, expansion of the durable goods industries overwhelmed growth in the services sector, reversing the shift away from manufacturing in the local economy. In contrast, manufacturing's share of employment in both Massachusetts and the nation continued to decline. Lowell's nonelectrical machinery industry, which includes the manufacture of computers and office machines, led the local employment boom, expanding an average of 43 percent a year from 1976 to 1982. The instruments industry grew by 24 percent a year during that period. In addition, employment in the transportation equipment industry, which is dominated locally by the guided missiles and space vehicles industry, expanded by 15 percent a year. Comparable rates of employment growth at the state and national levels were less than 7 percent in the nonelectrical machinery industry (less than 2 percent nationally) and less than 6 percent in the instruments industry, while there were actual declines in the transportation equipment industry. (See Table 5–1.)

By 1982, 39 percent of all nonagricultural employment in the Lowell area was in manufacturing, in contrast to only 24 percent in Massachusetts and 21 percent in the nation. (See Table 5–2.) More than one-quarter of all workers in Lowell were employed in the durable goods manufacturing sector alone.

High-technology industries account for a relatively large proportion of jobs in the Lowell economy. For instance, high-technology manufacturing employment accounted for 24

Table 5–1   Change in Employment by Industry for the Lowell Labor Market Area, Massachusetts, and the United States, 1976–1982.

Industry	Average Annual Change (%)		
	Lowell Labor Market Area	Massachusetts	United States
Construction	+0.3%	+ 2.8%	+1.6%
Manufacturing, total	+8.1	+ 1.2	−0.1
Nondurable goods, total	−3.3	− 1.5	−0.4
Food and kindred products	−3.8	− 2.7	−0.6
Textile mill products	−4.0	− 4.1	−3.1
Apparel and other textile products	−4.9	− 2.3	−2.0
Paper and allied products	+0.8	− 1.6	−0.9
Printing and publishing	−1.8	+ 1.8	+2.6
Chemicals and allied products	−1.6	− 0.8	+0.6
Leather	−8.4	− 3.8	−2.6
Other nondurable goods	+0.1	− 0.1	+1.0
Durable goods, total	+21.5	+ 3.1	+0.1
Lumber, wood and furniture	−3.6	N.A.	−1.3
Primary metals	+1.0	− 0.7	−3.4
Fabricated metals	+3.8	− 1.0	−0.8
Nonelectrical machinery	+43.3	+ 6.9	+1.6
Electrical and electronic equipment	+7.2	+ 5.5	+2.3
Transportation equipment	+14.7	− 0.2	−0.5
Instruments	+23.6	+ 5.2	+4.1
Miscellaneous manufacturing	−0.4	− 4.6	−1.5
Transportation, communication, and public utilities	+3.9	+ 0.9	+1.8
Wholesale and retail trade	+4.3	+ 1.6	+2.5
Finance, insurance, and real estate	+2.6	+ 3.7	+4.2
Services and mining	+6.8	+ 5.4	+5.3
Government	N.A.	− 0.4	+1.0
Total nonagricultural employment	N.A.	+ 2.1	+2.1
Total private, nonagricultural employment	+6.2	+2.6	+2.4

Source: Patricia Flynn, "Technological Change, the 'Training Cycle,' and Economic Development," from John Rees (ed.) *Technology, Regions and Policy* (Totowa, New Jersey: Rowman & Littlefield, 1986), p. 289. (reprinted with permission); Massachusetts Division of Employment Security, unpublished ES-202 data on the Lowell labor market area; U.S. Department of Labor, Bureau of Labor Statistics (BLS), *Employment and Earnings, States & Regions, 1939–82*, Vol. 1 (Washington, D.C.: U.S. Government Printing Office, 1983); U.S. Department of Labor, BLS, *Employment and Earnings, 1939–78* (Washington, D.C.: U.S. Government Printing Office, 1983); and U.S. Department of Labor, BLS, *Supplement to Employment and Earnings* (Washington, D.C.: U.S. Government Printing Office July, 1983).

N.A. = not available.

Table 5–2 Distribution of Nonagricultural Employment by Industry for the Lowell Labor Market Area, Massachusetts, and the United States, 1982.

Industry	Lowell Labor Market Area	Massachusetts	United States
Construction	3.3%	3.0%	4.4%
Manufacturing, total	39.2	24.3	21.0
Nondurable goods, total	11.4	8.4	8.7
Food and kindred products	1.1	0.9	1.8
Textile mill products	2.9	0.8	0.8
Apparel and other textile products	0.9	1.4	1.3
Paper and allied products	1.0	1.0	0.7
Printing and publishing	2.6	1.7	1.4
Chemicals and allied products	0.5	0.7	1.2
Leather	0.8	0.7	0.2
Other nondurable goods	1.5	1.1	1.1
Durable goods, total	27.7	15.9	12.4
Lumber, wood and furniture	0.3	0.4	1.2
Primary metals	0.7	0.6	1.0
Fabricated metals	0.9	1.8	1.6
Nonelectrical machinery	16.1	4.0	2.5
Electrical and electronic equipment	4.0	4.3	2.2
Transportation equipment	3.5	1.2	1.9
Instruments	1.8	2.3	0.8
Miscellaneous manufacturing	0.5	1.4	1.1
Transportation, communication, and public utilities	3.9	4.5	5.7
Wholesale and retail trade	21.6	21.7	22.8
Finance, insurance, and real estate	2.8	6.4	6.0
Services and mining	16.1	26.0	22.6
Government	13.2	14.0	17.6
Total nonagricultural employment	100.0	100.0	100.0

Source: Massachusetts Division of Employment Security, unpublished ES-202 data on the Lowell labor market area; U.S. Department of Labor, Bureau of Labor Statistics (BLS), *Employment and Earnings, States & Regions, 1939–82*, Vol. 1 (Washington, D.C.: U.S. Government Printing Office, 1983); U.S. Department of Labor, BLS, *Supplement to Employment and Earnings* (Washington, D.C.: U.S. Government Printing Office July, 1983).

Note: Columns may not add to 100.0% due to rounding.

percent of all employment in the Lowell area in 1982.[6] By comparison, such employment accounted for less than 10 percent of total employment in Massachusetts and less than 4 percent of all employment in the United States in 1982.

Employment in high-technology in the Lowell area has been concentrated in four industries: office, accounting, and computing machines; guided missiles and space vehicles; electrical and electronic equipment; and instruments. Ninety-seven percent of the growth in local high-technology industries between 1976 and 1982 occurred in these four industries. One industry in particular — office, accounting, and computing machines — dominates, generating over 70 percent of the high-technology employment growth between 1976 and 1982. By 1982, this industry accounted for 40 percent of all manufacturing employment in the Lowell area — 16 percent of all employment.

Employment by Occupation

The occupational composition of the high-technology industries entering the area differs significantly from that of Lowell's traditional manufacturing industries. For instance, in the high-technology industries vital to the Lowell transformation, a much higher proportion of employees are in professional and technical occupations than in the more traditional industries. In addition, the newer industries employ a relatively large percentage of skilled production workers. A substantially lower percentage of the work force in these high-technology industries is in low-skill and unskilled jobs.

More specifically, a comparison of the occupational composition of four traditional industries in which job loss in Lowell has been the greatest in recent years (apparel, leather, textiles, and food) with that of the four high-technology industries in which job growth has been concentrated (office, accounting, and computing machines; guided missiles and space vehicles; electrical and electronic equipment; and instruments), demonstrates the following:

• Professional workers account for less than 2 percent of the employment in the four traditional industries, but comprise between 9 percent and 26 percent of the workers in the four high-technology industries (see Table 5–3).

• Engineers make up 1 percent or less of the work force in the traditional industries, in contrast to 8 percent to 20 percent of the workers in the high-technology industries.

• Skilled technicians constitute less than 2 percent of the total work force in the traditional industries, whereas the range in the high-technology industries is from 6 percent to 11 percent.

• Low-skilled and unskilled operatives and laborers account for a majority of all jobs in the apparel, leather, and textiles industries, compared to between 19 and 36 percent of the jobs in the newer industries.

However, while professional and technical occupations comprise a greater share of jobs in the newer industries than in the traditional ones, blue-collar and clerical jobs continue to account for the majority of workers in the high-technology industries. In fact, unskilled assembler was the largest occupation in this group of local high-technology industries.

Earnings and Wages

The shift in employment to the high-technology industries raised earnings and wages in the Lowell area. Many of the newer firms in these industries pay well relative to the service-producing sectors and to the more traditional manufacturing industries in the local economy. For instance, in the durable goods manufacturing sector — which includes the high-technology industries — average annual earnings were approximately one-third higher than the average for all industries in the Lowell area in 1980.[7] In contrast, average annual earnings in the services and retail sectors that year were almost 25 percent and 40 percent, respectively, below the average for all local industries.

Within the better-paying, local high-technology sector, however, there is, substantial diversity with respect to earnings. In the nonelectrical machinery industry, average annual earnings were more than 40 percent above the average for all Lowell-area industries in 1980 — more than double those in the apparel industry, more than 75 percent above those in the leather industry, and more than 30 percent above those in textiles.[8] In the electrical and electronic equipment industry, by contrast, average annual earnings were 6 percent below the average for all industries and 13 percent below the average in textiles.

Data on wages paid to production workers further illustrate pay disparities within high-technology industries. In 1979 average hourly wages of production workers in the Lowell area were $4.71 an hour in the nonelectrical machinery industry and $6.17 an hour in the transportation equipment industry.[9] These wages compared favorably with those paid in textiles ($4.55 an hour) and in apparel ($4.28 an hour). Average wages of production workers in the electrical and electronic equipment industry, however, were $3.83 an hour — considerably below those paid in the declining traditional industries.

THE REINDUSTRIALIZATION PROCESS

How did the reindustrialization of this local economy occur? Why was Lowell the site of this high-technology "success story"? Specifying causality is complex, as many factors interacted and mutually reinforced each other. The characteristics of the local labor market, however, stand out: the Lowell area offered the mix of human resources desired by employers in the high-technology industries. In addition, the local economy provided these newer industries a competitive advantage.

The Human Resource Mix

The Lowell area offered both ready access to a good supply of highly skilled professional and technical workers and an abundance of low-cost labor. Located close to Boston, the Lowell labor market benefited from access to the Massachusetts Institute of Technology (MIT) as well as many other colleges and universities in the area that generate an ongoing supply of graduates in professional and technical disciplines. These institutions of higher education also serve as a source of new entrepreneurs.

In addition, being situated just outside Route 128, which by 1975 was densely populated with firms engaged in R&D and

Table 5–3   Occupational Distribution of Selected Traditional and High-Technology Manufacturing Industries in Massachusetts, 1980.

| | Traditional Industries | | | | High-Technology Industries | | | |
Occupation	Apparel	Leather	Food & Kindred Products	Textiles	Office, Accounting, & Computing Machines	Guided Missiles & Space Vehicles	Electrical & Electronic Equipment	Instruments
Managerial and professional specialty, total	5.7%	8.6%	11.9%	9.6%	31.4%	30.6%	20.9%	21.2%
Executive, administrative, and managerial	4.7	7.1	10.4	7.7	14.8	4.9	9.6	11.6
Professional specialty	1.0	1.5	1.5	1.9	16.6	25.7	11.3	9.6
Engineers	0.2	0.5	0.4	1.0	10.8	19.4	8.3	8.0
Technical, sales and administrative support, total	11.8	16.4	19.3	18.3	36.3	26.0	22.4	27.9
Technicians	1.0	0.5	1.0	1.7	11.1	8.6	6.6	9.8
Sales	1.5	2.5	4.4	1.5	1.9	0.0	1.9	2.2

Administrative support (including clerical)	9.3	13.4	13.9	15.1	23.3	17.4	13.9	15.9
Service	1.2	2.0	3.4	1.6	2.2	2.2	1.8	2.0
Farming, forestry, and fishing	0.0	0.0	0.0	0.0	0.0	0.0	0.0	0.0
Precision production, craft, and repair, total	9.3	10.6	20.7	16.3	11.1	13.7	18.6	18.6
Mechanics, repairers, and construction trades	1.3	1.7	4.7	8.0	3.9	4.9	4.0	5.4
Precision production	8.0	8.9	16.0	8.3	7.2	8.8	14.5	13.2
Operators, fabricators, and laborers	72.0	62.5	44.7	54.1	19.0	27.5	36.2	30.4

Source: Patricia Flynn, "Technological Change, the 'Training Cycle,' and Economic Development," from John Rees (ed.) *Technology, Regions and Policy* (Totowa, New Jersey: Rowman & Littlefield, 1986), pp. 290–91 (reprinted with permission); U.S. Department of Commerce, Bureau of the Census, *Detailed Characteristics of the Population: Massachusetts* (1980).

Note: Columns may not add to 100.0% due to rounding.

innovative activities, employers in the Lowell area were relatively close to a wealth of experienced scientific and engineering
talent. Skilled workers in those rapidly growing, high-technology firms had relatively high rates of turnover in the late 1970s
as favorable labor markets made "job hopping" lucrative. The
local economy also benefited from entrepreneurial "spinoffs," as
several former employees of established high-technology firms
along Route 128 chose to start ventures of their own in the
Lowell area.

In the 1960s and 1970s, the Lowell area also had an abundance
of relatively low-cost, lesser-skilled labor. Unemployment rates
in the area had been well above national norms for several
decades. Relative to the state and to the nation, production
wages in the Lowell area were low. In 1982, for instance, average
hourly wages of manufacturing production workers in the
Lowell area were 9 percent below the state average and 19
percent below the national average.[10]

In fact, during the period of revitalization, local production
wages fell on average relative to Massachusetts and the United
States. The increasing wage differentials resulted from the fact
that jobs in all three geographic areas expanded primarily in
those manufacturing industries in which Lowell's production
wages were low relative to state and national averages. Simultaneously, job declines were concentrated in industries in which
local employers were paying rates above state and national
averages.

In the late 1970s, for example, production workers in the
booming nonelectrical machinery and the electrical and electronic equipment industries in the Lowell area were earning less
than two-thirds the national average for production workers in
these industries. In contrast, local employers in the declining
textile and apparel industries were paying wages above the
national average for these more traditional industries.

Competitive Advantages of the New Employment

The high-technology industries that revitalized the Lowell
economy had a competitive advantage from both a national and

a local standpoint. From a national perspective, the area surrounding Route 128 provided the kinds of agglomeration economies that derive from an established high-technology base.[11] To summarize the earlier discussion, the region offered newly trained as well as experienced, highly skilled professional workers; a wealth of new, entrepreneurial talent; and an abundance of relatively low- cost production labor. In addition, the Lowell area had very few unions, further bolstering its attractiveness to high-technology employers.

From the local perspective, the Lowell economy provided a setting that was highly receptive to the jobs in these newer industries. Compared with other employment opportunities in the area, the jobs in the high-technology sector were widely regarded as better. Although the high-technology firms in Lowell were paying less than the national averages for production workers in these industries, these wages compared favorably to those paid in the more traditional sectors of the local economy.

The occupational composition of the high-technology industries also provided a more highly skilled mix of jobs than had previously been available in the area. As demonstrated earlier, the proportion of professional and technical jobs and highly skilled and semiskilled production jobs was higher in Lowell's high-technology sector than in its more traditional manufacturing industries. A substantially lower percentage of the work force in the newer industries was in low- and unskilled jobs than in the older industries.

In addition to relatively high wages and skilled jobs, modern facilities, favorable promotion prospects, and fringe benefits packages (which often included dental insurance, profit sharing, stock options, and pension plans) gave the rapidly growing high-technology firms an edge in both recruitment and retention of local workers. The local high-technology sector also enjoyed a very positive image—an image aggressively fostered by the local media.

Factors Accelerating the Transition

While the labor market conditions set the stage, an influx of private and public funds and strong local leadership accelerated the pace of the transition.

Millions of dollars from both the private and the public sectors fueled the local redevelopment effort. In the mid-1970s, for example, local bankers established the Lowell Development Finance Corporation (LDFC), committing 0.5 percent of their savings account funds to redevelopment projects in the city. The Lowell Plan was founded several years later to consolidate and coordinate development efforts and monies. This project— which included plans to renovate the local auditorium, increase local educational standards, and establish a first-class hotel in the downtown area—raised over $32 million in private funds. These funds were leveraged to secure over $74 million in matching state and federal dollars. In addition, millions of state dollars flowed into the local economy through Industrial Development Bonds, which provided low-interest loans to small firms to create or expand employment.

Substantial sums of federal money also flowed into the local economy in the late 1970s. The Lowell National Historical Park, the first urban park of its kind in the United States, was established. In addition, federal Urban Development Action Grants (UDAGs), which support firms developing projects in economically depressed areas, bolstered the reindustrialization.[12] For example, a UDAG loan worth $5 million encouraged Wang Laboratories, Inc., a producer of computers and office equipment, to build its worldwide headquarters in Lowell.

Strong local leadership also characterized the Lowell transformation. In particular, congressional influence, active participation of business leaders in community affairs, and innovative and aggressive public officials helped bring the local reindustrialization to fruition.[13]

STRUCTURAL CHANGE AND THE SKILL-TRAINING LIFE CYCLE

Aggregate trends in labor market data show the dramatic shifts in the industrial and occupational mix of employment during the Lowell reindustrialization. Interviews with local employers provide considerably greater detail on the human resource needs evolving from the rapidly changing structure of labor demands. They highlight the changing responsibilities of employers and schools in providing job-related skills as technologies evolve. In addition, they reveal the nature of skill shortages generated as the high-technology sector supplanted the more traditional employment base.

The Lowell example highlights how and why employers assume the responsibility for providing most job-related skills during the first phase of the skill-training life cycle. It also shows how educational institutions assume an increasing role in the provision of occupational skills during the intermediate phases of the cycle, and gradually relinquish this role back to individual firms.

Phase 1: New and Emerging Skills

Employers in the Lowell area provided the relatively firm-specific skills that emerge during periods of rapid product innovation. The high-technology firms engaged in R&D and made frequent design changes in their products. As a result, they experienced relatively short production runs. The manufacture of unique products or custom-designed components also involves what employers referred to as "hybrid assembly," which is done primarily by hand. Accordingly, considerable testing and inspection tasks were required.

Employers involved with these early stages of product development expressed a preference for "untrained, but trainable" production workers. Schools are not equipped to provide such training, as skills are often specific to the firm and acquired through on-the-job training.

The small subcontracting firms in the area, most of which are in the electrical and electronic equipment industry, generally performed relatively routinized production activities — such as cutting or assembling printed circuit boards — for large computer, office equipment, and military defense products. Production runs at these subcontracting firms were, however, also relatively brief, for board and assembly specifications varied by the product.

Phases 2 and 3:   The Shift of Training to the Schools

With growth in many of the high-technology firms, greater product standardization, and the adoption of specialized machinery for longer production lines, firms began seeking technically trained people from local educational institutions. In the early 1970s, very few occupational training programs were available in the immediate Lowell area. Hence several employers turned to private technical institutes, including Control Data Institute, Sylvania Technical Institute, and Microwave Training Institute, located outside the local labor market. Although expensive and somewhat inconvenient, this approach was preferable to in-house training, for these employers frequently needed only one or two workers with particular skills. Some employers maintained ongoing, direct contact with these private schools to insure that graduates would be referred to them. In other instances, employers paid for their current employees to attend these institutions.

The number of locally trained graduates with degrees in the technical fields increased in the late 1970s and early 1980s. According to the personnel director of one large computer firm, "Five years ago we recruited nationally, now local schools are giving us a good product." An executive of a much smaller firm, which produces transitors for electronic medical equipment, noted, "We used to promote and train from within, now we hire locally trained technicians from the vocational schools or the community colleges." In addition, an executive of a large computer manufacturer commented, "We now can hire technical people with related experience; we couldn't at first."

As computers became more widely adopted in offices as well as on the factory floor during the 1970s, local employers found it difficult to find graduates with computer-related training. A few employers also indicated nationwide skill shortages in specific engineering fields, such as microwave and software engineering. The most critical skill shortages during the reindustrialization process, however, appear to have been in technical fields. In addition, shortages of clerical and craft skills were cited by employers outside the high-technology sector.

Technical Skills

As just mentioned, shortages of technical skills occurred primarily in the late 1970s. Several of the larger computer and office equipment firms noted the need for technically trained workers to control and monitor newly installed equipment. One employer indicated, for example, that his firm introduced machines to make their own printed circuit boards—"doing work that a few years earlier was all done by hand." In addition, some firms installed computer-controlled equipment to assemble printed circuit boards. Skilled soldering and wiring tasks, formerly manual, were automated, and technical and computer-related skills were needed to program, setup, and maintain the equipment.

The adoption of new types of equipment into the more traditional manufacturing firms also generated demands for technical skills. In apparel firms, for example, new sewing machines automated positioning the needle, bar tacking, and trimming the material. Overall, fewer and lesser-skilled production workers were required to maintain output levels with the adoption of the new equipment. Several of the more difficult production tasks, however, still required skilled weavers and stitchers. Moreover, it was now essential to have some workers with technical skills who could keep the machinery in good running order. As one textile employer noted, "The machines are much more complicated, but the work isn't. We need far fewer workers, and most with lesser skills. However, we now have additional need for some technically oriented people who can troubleshoot. We can't afford to have any of the machines inoperable. In fact, preventive maintenance has taken on a whole new meaning."

In the early 1980s employers indicated that shortages were primarily for experienced workers. In particular, they found it hard to find college graduates with three to five years of experience for mid- and senior-level systems analysis, computer programming, and technical positions. One large computer firm awarded $500 bonuses to its workers for successfully recruiting such individuals.

Clerical and Craft Skills

Shortages of clerical skills—and to a lesser extent, craft skills such as carpentry and welding—occurred primarily outside the high-technology sector. Many employers in the more traditional manufacturing and service sectors found it difficult to hire skilled clerical workers. Employers in the high-technology industries, however, rarely indicated this problem. In fact, the latter employers generally did not have to advertise to fill clerical job openings; throughout the period they received a goodly number of unsolicited résumés from skilled clerical workers, many of whom had years of experience.

The uneven impacts of the clerical and craft skill shortages, attributed to the rapid growth of the local high-technology sector, were generally not attributed to wage rates. Many of the firms in the more traditional sectors paid these clerical and craftsworkers rates comparable with the high-technology firms. Employers in the more traditional sectors indicated, however, that they were at a competitive disadvantage in terms of fringe benefits packages and overall "image."

Skill Shortages and the Small Firm

Skill shortages varied according to firm size. Small firms, in particular, were at a disadvantage in hiring and retaining workers with skills in new and emerging fields. Limited training budgets and smaller pools from which to draw internal candidates made the smaller firms more dependent than large firms on external sources for skilled workers. During the early phase of skill development, however, schools generally are unable to provide training in these fields.

Several of the relatively small firms lowered their entry requirements and provided the necessary training on the job. Even so, they often lost these trained workers to larger firms that offered higher wages and greater promotion opportunities. Personnel officers at the larger firms indicated a preference for hiring "experienced" workers at all levels, even if the new recruits' skills were not relevant to the job for which they were being hired. "We're like a training house," noted one small subcontracting employer that regularly loses its workers to its relatively large customers. "[We are] better off lowering entry standards and providing 'targeted training' in-house to novices than raising everyone's wages."[14]

Labor Bottlenecks

By 1984 a general labor shortage throughout the Lowell area contributed to skill shortages. In the 1960s and 1970s, employers were able to entice local residents who were not already in the labor market, particularly women, to come to work. By 1980, however, labor force participation rates of both men and women were considerably above the national averages.

Low-skilled and unskilled immigrants and refugees constitute a relatively recent source of new workers in the Lowell area. In particular, Cambodians, Laotians, and Vietnamese, many of whom do not speak English, are being hired in both the traditional and high-technology sectors.[15]

The growing mix of nontraditional workers with relatively low levels of education and work experience will generate further challenges to employers and schools in seeking to meet local skill needs.

Phase 4:   Skilled Replacement Needs for Mature Products

The high-technology firms in Lowell have not yet reached mature stages of production, but we can look at the experiences of firms in the traditional textile, shoe, and apparel industries to obtain insight into what can be expected in phase 4 of the skill-training cycle.

Local firms with relatively mature product lines and stable or declining demands provided their own training to meet skilled replacement needs.[16] In interviews, these employers indicated that their relatively older work forces had been retiring at high rates in recent years. Hiring trained workers to fill these skilled positions, however, was virtually impossible. Schools in the area offered little, if any, job-related training for these needs — labor demands in these occupations had been dwindling for years.

Moreover, recent attempts to establish training programs in these fields proved unsuccessful. Although shoe and apparel firms donated time, money, equipment, and instructors to the schools to run programs, and local employers guaranteed graduates jobs upon program completion, the programs failed to attract enrollees.[17]

School officials indicated that students, and their parents, view training for the remaining traditional manufacturing firms in the area as highly risky, and at best a short-run employment strategy. Moreover, many employers in the traditional manufacturing firms indicated that they have discouraged their own children from entering these fields. Noting it to be both an expensive and time-consuming proposition, these employers said that they have lowered their entry standards — the ability to speak or read English, for instance, is no longer required — and provide the training on the job.

SUMMARY

The transformation of the local Lowell economy as high-technology industries supplanted the more traditional employment base triggered dramatic structural changes in labor demands. The skill and occupational needs of the newer firms were wide-ranging and rapidly evolving. Skill shortages, particularly in technical areas, occurred as labor demands in the newer firms outpaced the supply of trained graduates. Moreover, requirements for clerical and craft skills of the firms in the rapidly growing high-technology industries exhibited spillover effects into the more traditional employment sectors.

The responsibility for providing training for the needs of the newer industries shifted from the employers to the schools as labor demands rose and skills became more transferable among workplaces. Data from employer interviews help to further delineate their needs and concerns over the phases of the skill-training life cycle. The following chapter looks in more detail at the role of various educational institutions in this skill-transfer process.

6 EDUCATION AND CHANGING SKILL NEEDS

Educational institutions providing job-related training can facilitate technological change in local economies by adapting to the diverse and evolving skill needs of an area. Monitoring growth and decline over the course of industrial development, however, is not easy. Of particular difficulty is anticipating new and emerging skill needs.

The reindustrialization of Lowell illustrates how educational institutions can assist an economy experiencing rapid shifts in the structure of labor demands.[1] The Lowell example also provides the chance to further explore skill-training life cycles as technologies evolve.

As discussed in Chapter 5, the occupational composition of the high-technology industries critical to Lowell's revitalization encompassed a much greater proportion of highly skilled and semiskilled jobs than the more traditional industries in the area. These high-technology jobs also tended to pay well, relative to the jobs in older industries in the local economy.

Highly skilled engineers and scientists were available from the nearby Boston and Route 128 areas. However, when the Lowell economy began to turn around after three decades of stagnation, very few occupational training programs were available in the local area. During the subsequent period of rapid growth and industrial change, the local occupational education network evolved in ways highly reflective of labor market trends. This chapter provides a detailed analysis of the response of the local schools and colleges to the changing skill requirements, highlighting the particular needs of the high-technology industries. The key factors stimulating the overall educational response are identified, as are those constraining change. Institutional and programmatic differences also are evaluated.

Using data for specific institutions and programs for the 1970–1982 period, as well as information acquired in interviews with educators, employers, and government officials, this chapter analyzes the supply of occupationally educated graduates in light of local labor market trends. Placement data in the Lowell area confirm the finding of other studies: that most graduates of programs below the baccalaureate level acquire their first full-time jobs relatively close to home.[2] The supply of graduates from local four-year college and graduate degree programs is analyzed separately, from a broader geographic perspective, as many of these individuals compete in regional or national labor markets for their first jobs after graduation. (See Appendix 6A.)

Analysis of the occupational education data by institution and by educational program permits delineation and a better understanding of the dissimilar roles of educational institutions in providing skills for particular occupations and in responding to the changing labor market needs. This level of disaggregation permits study of the interaction between particular firms and educational institutions, and demonstrates the interrelatedness of the various components of the occupational training network.

OVERVIEW OF THE LOCAL OCCUPATIONAL EDUCATION NETWORK

The educational institutions providing occupational education in the Lowell area during the time of the study include one public university, two public four-year colleges, one private hospital-affiliated school of nursing, two public two-year colleges, one private two-year college, one citywide public vocational school, three public regional vocational schools, three proprietary schools, and eight local high schools. (Throughout this chapter the term *vocational education* refers only to programs offered at the four vocational schools; the term *occupational education* is used to describe all programs providing job-related skill training below the four-year college degree level.)

Over 25,000 Lowell-area graduates from programs below the four-year college degree level entered the labor market between 1970 and 1982.[3] Of these, over 40 percent received training from some clearly defined skills programs. (See Table 6–1.) Vocational schools accounted for just over half these trained graduates; two-year colleges, for 22 percent; associate degree programs at four-year colleges and universities, 16 percent; proprietary schools, 7 percent, and private, nonprofit institutions, 3 percent. The vast majority (89 percent) of these graduates were trained at public-sector institutions, and just over half the graduates received their training at the secondary level.

The skill training of the remaining 15,000 graduates cannot be determined precisely, as they are local high school graduates who were not classified by "track." Some of these graduates clearly acquired job-related skills, whereas others did not. Students taking business courses at the local high schools, for instance, were more likely to obtain "marketable" skills than were graduates who had taken college-preparatory courses. In addition, some high school students took industrial arts courses, such as electronics, general shop, and power mechanics. Only one or two courses are offered at the local high schools in each of these fields, so that the training is much less intensive than that offered in the regional vocational schools.

Because their occupational training was indeterminable, the graduates from the local high schools are excluded from the following statistical analysis of occupationally trained graduates. These schools are included thereafter, however, in the analysis of institutional change.

TRENDS IN OCCUPATIONAL TRAINING

The number of occupationally trained graduates as well as the variety of fields in which they were trained grew substantially over the 1970–1982 period. (See Table 6–2.) The total number of graduates increased at an average annual rate of 17 percent. By broad occupational training field, graduates of office programs

Table 6–1   Potential Labor Market Entrant Pool of
Occupationally Trained Graduates in the Lowell
Labor Market Area, 1970–1982.[a]

School	Number of Graduates Not Going on to Further Education	Percent of Total Occupational Education Graduates
Vocational schools, total	5,353	51.8%
Lowell Trade	678	6.6
Greater Lowell Regional Voc. Tech.	3,159	30.6
Nashoba Valley Tech.	926	9.0
Shawseen Valley Voc. Tech.	590	5.7
Community colleges, total	2,186	21.2
Middlesex Community College	1,083	10.5
Northern Essex Community College	1,103	10.7
Private two-year colleges		
Newbury Junior College	72	0.7
Public four-year colleges, total	1,629	15.8
Lowell Technical Institute	982	9.5
University of Lowell	647	6.3
Proprietary schools, total	739	7.2
Lowell Academy of Hairdressing	564	5.5
Solari's School of Hairdressing	175	1.7
Private nonprofit institutions		
Lowell General Hospital School of Nursing	352	3.4
Total	10,331	100.0%

Note: This table, generated from school records and follow-up surveys, estimates the "potential" labor market entrant pool from these training programs. That is, these labor market entrants include any student who did not intend to enter a postsecondary educational program in the fall after graduation. Percentages may not add to 100.0% due to rounding.

[a]Below the four-year college degree level. Table does not include graduates from the local high schools. Numbers of graduates are adjusted to account for dissimilar educational service areas and the labor market area.

showed the greatest growth (+38 percent a year), while graduates in trade and industry and in health also grew at above-average annual rates, 22 percent and 18 percent, respectively. In contrast, the number of graduates of the consumer and home-making field grew relatively slowly, at 7 percent a year on average, while the number of technical graduates actually declined slightly. Distributive education and public services represented new fields of training introduced during this period.

Within several of these broadly defined fields of training, occupational offerings became more diverse. In the trade and industry area, for example, new programs added to the curricula include auto body; commercial art; commercial photography; electronics; graphic arts; heating, ventilation, and air conditioning (HVAC); masonry; mechanics (except automobile); metal fabrication; plumbing; radio and television; sheetmetal; and welding. Similarly, in the health field, programs in dental assistance, dental laboratory technology, dental hygiene, diagnostic medical sonography, gerontology, medical aide, medical records, mental health, and respiratory therapy were added to the original nursing and radiological health technology offerings.

In addition, educational offerings in other occupational fields were reorganized. In the consumer and homemaking area, for example, general home economics programs were replaced with more specialized and job-oriented food management and clothing management programs. In the technical field, electronics programs superseded electrical programs.

LABOR MARKET RESPONSIVENESS

The occupational education network was highly responsive to overall occupational trends in the area and to the particular needs of the high-technology industries. Three-quarters of the occupational education programs, accounting for 85 percent of all of the trained graduates, were "on target" or "reasonably aligned" with occupational employment changes in the Lowell area during the 1970s. (See Appendix 6B for methodology and details of the analysis.) Less than 3 percent of the graduates

Table 6–2   Occupationally Trained Graduates[a] in the Lowell Labor Market Area, by Program[b] 1970–1982.

Program	Year													Total
	1970	1971	1972	1973	1974	1975	1976	1977	1978	1979	1980	1981	1982	
Agriculture											1	0	1	2
Horticulture											1	0	1	2
Distribution education			1	5	7	7	4	6	11	16	14	16	28	115
Health, total	64	59	75	105	115	122	151	174	182	198	179	176	198	1,798
Dental assistant			3	6	6	5	5	7	6	7	5	3	13	66
Dental hygienist					7	7	6	9	7	7	11	10	11	75
Health administration					1	6	11	13	11	8	5	—	—	55
Nursing	59	56	63	77	67	73	91	91	85	91	88	96	96	1,033
Medical aide				4	10	15	22	35	29	44	38	29	38	264
Medical technician	5	1	6	16	18	15	16	19	26	24	21	22	21	210
Medical records		2	3	2	6	5	8	10	12	9	6	4	5	72
Other, health									14	13	9	12	14	62
Consumer and homemaking, total	32	37	26	42	14	24	47	64	69	64	62	80	57	617
Child care				3	8	16	22	21	25	18	14	24	21	172
Food management				3	4	5	20	33	29	38	34	46	29	241

														Total
Clothing management								9	6	6	13	6	5	45
Home economics	32	37	26	36	2	3	5	1	9	2	1	3	2	159
Office, total	43	47	96	111	145	125	150	108	209	210	202	207	239	1,892
Accounting	4	4	2	3	4	4	5	3	13	9	7	18	23	99
Data processing	3	4	6	2	5	7	5	7	29	26	32	40	45	211
Supervisory/mgt.	26	32	62	67	85	67	74	39	36	40	47	50	66	691
Secretarial/clerical	10	7	26	39	50	46	65	54	124	125	114	96	98	854
Other office					1	1	1	5	7	10	2	3	7	37
Technical, total (excluding health)	101	109	94	133	116	147	88	58	62	58	32	65	94	1,157
Civil eng. tech.	36	26	40	41	31	45	20	8	6	7	3	6	9	278
Computer tech.	3	2	2	4	3	2	4	3	8	8	8	14	19	80
Electrical tech.	23	32	26	26	20	29	–	–	–	–	–	–	–	156
Electronic tech.	2	3	2	2	2	1	25	14	12	15	6	15	33	132
Electromechanical tech.									3	2	4	4	5	18
Mechanical tech.	14	21	10	10	8	11	11	14	13	10	2	15	12	151
Industrial tech.	6	5	5	5	5	7	8	7	2	5	0	1	0	56
Other tech.	17	20	9	45	47	52	20	12	18	11	9	10	16	286

(continued)

Table 6-2 (continued).

Program	1970	1971	1972	1973	1974	1975	1976	1977	1978	1979	1980	1981	1982	Total
Trade and Industry, total	151	142	213	229	249	277	408	472	633	643	555	501	540	5,013
Air conditioning/ HVAC									21	24	16	19	18	98
Auto body			1	3	4	6	10	10	27	34	29	21	28	173
Auto mechanics/ diesel	21	11	26	27	28	37	45	58	54	58	51	46	40	502
Carpentry	22	18	34	47	26	42	53	59	65	55	47	42	33	543
Commercial art			6	4	9	6	12	9	15	31	23	25	21	161
Commercial photography								3	2	2	1			8
Cosmetology	59	67	53	56	47	48	52	93	88	88	84	66	85	886
Drafting	7	5	15	7	16	20	13	30	42	41	32	54	37	319
Electricity	15	21	26	28	39	35	59	50	42	37	34	17	40	443
Electronics			6	9	18	20	26	27	46	32	28	20	45	277
Graphic arts			4	4	6	7	18	28	42	43	38	31	21	242
Machine shop	19	14	21	19	33	21	41	36	41	34	25	36	33	373
Masonry							1	2	23	17	22	15	13	93
Mechanics, excluding auto							1	0	12	22	17	15	11	78
Metal fabrication			6	10	6	9	13	13	9	10	12	4	3	95
Painting and decorating	3	5	9	8	4	14	23	16	29	28	20	31	23	213

														Total
Plumbing				3	8	6	26	24	31	33	18	22	27	198
Radio and T.V.										16	13	4	7	40
Sheetmetal									12	13	9	8	10	52
Upholstery	5	1	6	4	5	6	8	2	17	6	15	2	14	91
Welding							7	12	15	19	21	23	31	128
Public service, total			8	16	27	20	87	59	48	46	33	37	28	409
Law enforcement				2	9	7	19	22	31	28	23	21	10	172
Fire protection							2	4	7	5	7	4	5	34
Other public service			8	14	18	13	66	33	10	13	3	12	13	203
Total occupationally trained graduates	391	394	513	641	673	722	935	941	1,214	1,235	1,078	1,081	1,185	11,003

aBelow the four-year college degree level. Numbers of graduates are adjusted to account for dissimilar educational service areas and the labor market area.

b. 0 indicates that no one graduated from the program; — indicates that the program no longer existed.

were from programs experiencing slow growth or decline, although their related occupations were growing relatively rapidly. In addition, approximately 12 percent of all graduates were in programs that had grown rapidly in spite of below-average growth or declines in employment in those occupations.

Analysis of these latter training areas with "potential surpluses" shows that the programs were responsive to past employment trends in the area. More specifically, in eight of the nine cases involved, programs had been originally introduced into the local area after extensive expansion of employment in the 1960s in their related occupations. In each of these instances, however, much slower growth or decline occurred in the 1970s. Further analysis shows that annual replacement needs in these occupations were insufficient to absorb the number of new graduates—which was less than twenty-five per program per year. Students thus faced considerably weaker labor demand at graduation than expected.

In terms of meeting current employment needs, the local occupational education network performed better than if either past employment trends or occupational projections alone had guided the changes in the program offerings. For example, had past employment trends alone been used to direct educational change—that is, if training programs had been eliminated in areas in which past employment had declined, and new programs had been introduced in areas that had experienced relatively fast growth—a smaller percentage of the graduates (66 percent versus 85 percent) would have received training in programs that were "on target" or "reasonably aligned" with employment trends in the 1970s. In particular, rapid employment growth in the local area in the 1960s might have lead to expansion in programs in auto mechanics, cosmetology, nursing, civil technology, mechanical technology, and industrial technology. In fact, these programs experienced below-average growth or declines in the number of graduates, which was consistent with employment trends in their related occupations in the 1970s.

Had occupational projections been the sole determinant of educational change, many more local labor market demands would have gone unmet by this training system than was the actual case.[4] More specifically, if new programs had been introduced only for occupations projected to have at least average growth in the 1970s, there would have been no locally trained graduates in commercial art, computer technology, electromechanical technology, fire protection, law enforcement, graphic arts, HVAC, mechanics (except automobile), radio and television, and welding. In contrast to what was projected, however, the occupations related to each of these programs—which were, in fact, initiated during this period—experienced average or above-average growth in the local labor market during the 1970s.[5]

While the overall response of the occupational education network to labor market trends was good, some potential skill mismatches did occur. These were primarily in occupations for which training programs were introduced following a spurt of employment growth that failed to be sustained. The sudden turnaround or slackening of growth in these occupations caused graduates to enter labor markets with far fewer job opportunities than expected upon enrollment. Moreover, several other programs might have experienced a similar fate had a series of institutional constraints—such as a lack of equipment, physical plant, or funds—not hindered their growth. Many of these programs subsequently appeared "on target" when employment growth in these fields subsided.

Response to the Needs of the High-Technology Industries

Educational institutions in the Lowell area were particularly adaptive to the skill requirements of the booming high-technology sector. Occupational education programs providing skills needed by the local high-technology industries grew considerably more quickly than other training programs did. In addition, a relatively large proportion of the vocational education funds available to local institutions went toward providing skills for high-technology employers.

Whereas the relatively high proportion of scientific and engineering staff for which the high-technology sector is noted come from bachelor's and graduate programs, the much larger group of technical, precision craft, and administrative support employees generally receive training below the four-year college level. The difficulty of reaching consensus on the definition of "high technology" compounds the perennial problem in occupational education of "matching" particular training programs to specific occupations. One can, however, eliminate certain programs that appear to have little relevance to the high-technology sector. Furthermore, given the occupational composition of specific "high-technology" industries in the local economy, one can isolate programs most likely to serve as a source of skilled labor to these employers.

As noted in Chapter 5, four high-technology industries were responsible for the bulk of the new jobs in the Lowell area during the 1970s: office, accounting and computing machines, guided missiles and space vehicles, electrical and electronic equipment, and instruments. Based on a list of the twenty largest occupations in the state for these industries, in conjunction with placement records from the local schools and colleges, twelve of the occupational education network's forty-nine fields of training were identified as providing skills required by the local high-technology sector. These "high-technology education programs" include three in the office category — accounting, data processing, and secretarial; four in the technical category — computer technology, electrical technology, electronic technology, and electromechanical technology; and five in the trade and industry category — drafting, electricity, electronics, machine shop, and welding.[6]

From 1970 to 1982, the number of graduates from these high-technology education programs grew more than twice as rapidly as the number of graduates from the other occupational education programs. (See Table 6–3.) Moreover, during the latter half of this period, the supply of graduates from the high-technology programs grew four times more quickly than the supply of graduates from all the other occupational education programs.

The high-technology education programs in the office category experienced the fastest growth throughout the period. From 1970 to 1976 this growth was concentrated in the secretarial programs, while in the latter half of the period, growth in the data processing and accounting programs outpaced that in the secretarial areas. In the trade and industry category, growth

Table 6-3   High-Technology Occupational Education Programs in the Lowell Labor Market Area, 1970–1982.

Program[a]	Total Graduates, 1970–1982[b]	Average Annual Change in Graduates, 1970–1976 (%)	Average Annual Change in Graduates, 1976–1982 (%)	Average Annual Change in Graduates, 1970–1982 (%)
Technical, total	672	+4.2%	+3.9%	+9.0%
Computer tech.	80	+5.6	+62.5	+44.4
Electrical tech.	156	−100.0	0.0	−100.0
Electronic tech.	132	+191.7	+5.3	+129.2
Electromechanical tech.	18	–	new program	new program
Other tech.[c]	286	+2.9	−3.3	−0.5
Trade and Industry, total	1,540	+42.7	+4.6	+20.3
Drafting	319	+14.3	+30.8	+35.7
Electricity	443	+48.9	−5.4	+13.9
Electronics	277	new program	+12.2	new program
Machine shop	373	+19.3	−3.3	+12.3
Welding	128	new program	+57.1	new program
Office, total	1,164	+56.9	+22.2	+49.0
Accounting	99	+4.2	+60.0	+39.6
Data processing	211	+11.1	+133.3	+116.7
Secretarial	854	+91.7	+8.5	+36.1
High-tech programs, total	3,376	+31.7%	+9.6%	+29.7%
Non-high-tech programs, total	7,627	+20.5%	+2.4%	+12.9%
All occupational education programs	11,003	+23.2%	+4.5%	+16.9%

[a]Selected on the basis of the largest occupations in the high-technology industries that accounted for the bulk of the new jobs in the Lowell labor market area in conjunction with placement records from the local schools and colleges. (See text.)

[b]Adjusted to account for dissimilar educational service areas and the labor market area.

[c]Includes engineering science, applied chemistry, applied mathematics, plastics technology, technology, and environmental technology.

Source: Patricia Flynn, "Technological Change, the 'Training Cycle,' and Economic Development," from John Rees (ed.) *Technology, Regions and Policy* (Totowa, New Jersey: Rowman & Littlefield, 1986), p. 295.

shifted from the traditional electrical to the more high-technology-oriented electronics programs. A similar trend occurred in the technical category, as students moved out of the electrical technician programs and into the electronic technician and computer technician programs. Hence, although the total number of graduates in the technical fields declined—a pattern seemingly unresponsive to the rapid growth of high-technology jobs—the change in "mix" of the technically educated graduates clearly was consistent with those labor market trends.

By 1982, graduates from the high-technology training programs accounted for 36 percent of all occupationally trained graduates in this local economy, compared to 24 percent in 1970.

Less Formalized High-Technology Education Programs

The preceding analysis, based on data of graduates receiving an associate's degree, certificate, or diploma upon completion of the occupational education program, fails to account for a wide range of less traditional and less standardized education programs, many of which are in experimental stages. Yet these types of programs usually play a key role in providing skill training to "emerging" or "high-technology" occupations.

Between 1979 and 1982, federal vocational education funds supported over fifty high-technology programs in the schools and colleges in the Lowell area—programs that are not included in the preceding analysis (see Table 6–4). These additional high-technology programs, which ran from fifteen weeks to two years in length, cost over $1.5 million and accounted for almost half of all federal vocational education funds dispersed in the Lowell area during those years.[7] Over 70 percent of these federal vocational education funds were used to implement office training programs, including, for example, technical writing, word processing, and data processing. Trade and industry programs, such as printed circuit board design and electronics, accounted for 17 percent of these federal funds, while technical programs—including electronic technician training and computer maintenance technology—accounted for the remaining 12 percent of these monies.

Table 6–4   Federally Funded High-Technology Programs in the Lowell Labor Market Area, by Institution, 1979–1982.[a]

Project	Amount of Funding	Number of Individuals to be Served
Middlesex Community College		
Electronic Technician—Computer Option	$69,392	40
Training to Launch a Career (Including Office Skills and Word Processing)	40,225	75
Technical Writing	43,304	20
Drafting and Printed Circuit Board Design	25,542	20
Training to Launch a Career II	29,994	50
Training to Launch a Career III	40,225	20
Technical Writing II	32,572	20
Computer Operation	34,951	25
Technical Writing III	35,396	20
Northern Essex Community College		
Computer Maintenance Technology	$30,000	38
Computer Maintenance Technology II	8,069	38
Word Processing Technology and Management	22,900	25
Intensive Office Skills Training Program	23,500	25
Computer Service Repair	35,995	24
Greater Lowell Regional Vocational Technical School		
Graduates Employed as Technicians	$68,963	24
Graduates Employed as Technicians II	36,000	24
Word Processing	27,640	186
Retraining and Instruction Geared to High Technology	54,900	34
Graduates Employed as Technicians III	15,000	24
Skills Training Alternatives for Youth (Electronics and Computers)	14,479	15
Retraining and Instruction Geared to High Technology	57,591	38
Shawseen Valley Vocational High School		
Project Summer (Computer Programming and Word Processing)	$17,700	4
Project P.A.T. (Word Processing)	17,846	20
E.T.C.O. (Electronic Technician and Computer Operators)	16,450	36
DEC-20 (Computer Operators)	8,000	14
Computer Application Programmer	72,618	34
Homemakers Assessment Training (Office skills, including Word Processing)	20,096	33

(continued)

Table 6–4   (continued).

Project	Amount of Funding	Number of Individuals to be Served
Shawseen Valley Vocational High School (continued)		
Maintenance Mechanic (including Welding and Machine Shop training)	22,172	24
Computer Operator and Programmer	54,790	36
Computer Aided Training (Computer Programmers and Word Processors)	56,864	44
Billerica High School		
Office Simulation (Electronic Data Processing and Word Processing)	$25,453	39
Office Simulation II	16,982	53
A.S.T.O.P. (Data Processing and Word Processing)	9,786	54
Electronic Test Technician	12,560	15
Project Entry (Data Processing)	9,310	112
Project Transcribe (Office Skills)	12,164	40
Electronic Test Technician	17,361	20
Chelmsford High School		
Introduction to Computer Data Based Accounting	$18,263	177
Retraining Programs for Updating Skills (Office Skills)	38,167	100
Word Processing	22,547	51
Office Occupational Skill Training	32,538	164
Dracut High School		
Communications Technicians Training (Electronic Technology)	$15,060	32
Work Experience Program in Word Processing	29,367	55
Groton-Dunstable Regional High School		
Word Processing and Computer-Assisted Accounting	$ 4,359	40
Lowell High School		
Occupational Training—Word Processing	$25,000	200
Computer-Oriented Skill Program (Data Processing and Computer Programming)	82,378	80
Word Processing Instruction	12,881	30
Tewksbury High School		
Word Processing	$12,847	80
Project P.O.K.E. (Computer Programming)	16,740	28
Electronics	29,141	24

(continued)

Table 6–4   (continued).

Project	Amount of Funding	Number of Individuals to be Served
Tyngsboro High School		
Accounting and Computing	$ 3,973	118
Electronic Data Systems Management Laboratory	17,755	118
Westford Academy		
Business Update (Computer and Electronic Equipment	$ 8,291	95
Business Update II	8,291	95
Advanced Accounting – Microcomputer	11,032	130

Source: Unpublished records, Massachusetts Department of Education, Division of Occupational Education, 1979–1982.

a. Funded under the Vocational Education Act, as amended, Public Law 94–482. Funding amounts and numbers of individuals to be served are listed as stated on the approved budget request.

Federal funds were spent also on high-technology programs in local educational institutions through the Lowell Comprehensive Employment and Training Administration (CETA). Between 1979 and 1982 over $500,000 in CETA funds supported training in data processing, word processing, welding, drafting, and electronic technology. An additional half million dollars in CETA funds was appropriated for electronics programs at some of the local high schools.

Other high-technology education programs were supported by the state-funded Bay State Skills Corporation, an agency established in 1981 to provide matching grants to schools and colleges that work with private-sector firms to train people in high-growth occupations. In 1982 this quasi-public organization funded programs for electronic testers, electronic technicians, assembly machine operators, and plastics technicians in the Lowell area. The Bay State Skills Corporation also funded a program in advanced automation and robotics, including payment toward development of a full robotics program and laboratory, at the University of Lowell.

CHARACTERISTICS OF EDUCATIONAL RESPONSE

The occupational education network in the Lowell area clearly changed in directions reflective of contemporary labor market trends—even though these trends often were not discernible from past employment data nor occupational projections. Three factors were crucial to this educational response: the opening of several institutions providing occupational education programs to the area; the use of current employment trends to guide curricular change; and the use of federal vocational education funds to initiate programs for new and emerging skills.

New Institutions

As noted earlier, very few occupational education programs existed in the Lowell area when the economy turned around in the 1960s. More specifically, the occupational education network consisted of one citywide vocational school, one technical institute, and one hospital-affiliated school of nursing. From 1970 to 1982, three new regional vocational schools and one new community college opened in the area. These public institutions resulted in a threefold increase in the number of occupationally trained graduates. Moreover, their addition to the educational network more than doubled the number of occupational fields for which skill training was available locally. Regional labor market analyses and assessments of preexisting programs in the local area played key roles in determining the original curricula for these institutions.

Curricular Change

There was general consensus in the Lowell occupational education network that current local labor market needs should guide program development and curricular change. *Future* labor market needs were considered the ideal variable to guide such change; however, most local educators believed these to be "unknowable." Given the rapidly changing economy, local educators perceived little value in past employment statistics as predictors of future skill needs. There was also widespread reluctance to trust occupational employment projections. Several educators referred to the unpredicted, large-scale layoffs of skilled workers, particularly in the electronics fields, along Route 128 in the early 1970s.[8] Moreover, they noted that employment projections in the early 1970s had not foretold the rapid growth in high-technology industries that the area experienced shortly thereafter.

These educators view local employers as the key source of contemporary labor market needs. Employer input, obtained primarily via participation on advisory committees for various education programs or via cooperative education programs (in which students receive on-the-job training at the worksite to supplement their classroom instruction) often guided the content of new training programs. Employers were found particularly helpful in indicating current skill shortages, weaknesses in particular training programs and curricula, and problems experienced with recently hired graduates. In addition, they often provided guidance with respect to the types of training they sought from the schools' programs as opposed to training they were willing to provide at the workplace. Educators indicate, however, that local employers generally were *not* a good source of future labor market needs, as they rarely were able to define even their own firms' hiring requirements beyond a few months.

In addition to employer input, job placement rates, which in general were high throughout this occupational education network, were closely monitored for signals that training was not in line with employment needs. Educators cited difficulties, however, in distinguishing short-term, cyclical cutbacks in demand for longer-term structural changes. Placement rates alone also fail to indicate the extent to which placement problems are due to inappropriate occupational skills as opposed to employer ignorance of the training being provided.

Student demand was also an important consideration in institutional change, particularly in the decision to eliminate a program. Almost all the programs that were phased out in this

occupational education network had experienced low enrollments. The vocational schools had more leeway in this regard
than did the postsecondary institutions that had less "captive"
audiences. Programs run during the regular day sessions at the
regional vocational schools were at the secondary level: the
students' alternative was to attend their local high school. All
three vocational schools in this study had more applicants than
slots available, and very few students chose to return to their
high school. Enrollment quotas, determined in large part by
equipment availability, applied to most of the programs at the
vocational schools. When program quotas were met, students
were allocated to the less popular fields.

Federal Vocational Education Funds for New and Emerging Skill Programs

Federal vocational education monies were the major source of
"venture capital" for innovative programs designed to meet
new and emerging employment opportunities in the Lowell
area. As noted earlier, $1.5 million in federal vocational educa-
tion funds supported over fifty high-technology education programs in the area between 1979 and 1982. Approximately one-
third of these funds went to the community colleges, one-third
to the regional vocational schools, and the remaining third to
the local high schools. Each of these types of institutions spent
the majority of their federal vocational education funds on office
skill training. The community colleges spent their remaining
federal vocational education monies on technical training programs, whereas the vocational schools and high schools allocated the rest to trade and industry programs.

Adult students, many of whom were being retrained, were
the primary beneficiaries of these federally financed high-technology programs at the vocational schools and community
colleges. The programs at the local high schools were mainly for
secondary-level students. For the high-technology programs
offered at the local high schools and those in the technical fields
provided by the community colleges, the bulk of these federal
dollars were used to purchase equipment needed for the occupational training.

Constraints to Educational Change

Although changes in the occupational education network were highly reflective of local labor market changes, educators frequently cited the lack of up-to-date equipment as a hindrance to updating old programs and starting new ones in emerging fields. Rarely did any of these schools or colleges receive equipment donations from the business community. Trade and technical programs in particular require considerable investment in equipment, purchases of which are often amortized over eight- to ten-year periods. Equipment availability mandates quotas in almost all the trade and industry programs at the vocational schools. Equipment needs also constrained growth in the technical programs at the community colleges and in electronics and office skill offerings at the local high schools. Although previously considered a low-cost type of training, office skills programs are requiring greater and greater capital investments in computers, word processors, and the like.

Physical plant constraints also hindered growth of vocational school training. While two of the three vocational schools expanded their physical facilities during the period, applications continued to exceed available training slots.

Institutional Patterns and Trends over Time

Occupational education, by definition, is distinguished from other types of education by its specific goals of providing graduates with marketable job skills. But even though these institutions were operating under similar labor market conditions, they took on different roles according to their particular "missions" and to the availability of funds and equipment.

The regional vocational schools in this labor market provided the bulk of the training in the distributive education and in the trade and industry fields. The bulk of the programs, which offer secondary-level training over a four-year period, provide students a cluster of job skills as well as a strong theoretical foundation in the basic principles underlying the mechanical aspects of the skills. These students also take a full academic program. In addition, many of the vocational students participate in a cooperative education program, which provides supervised work experience related to their classroom instruction.

The community colleges were the primary source of skill training in public services and health. During this period the community colleges absorbed the bulk of the nursing and health technician programs, previously provided by the three-year nursing school and four-year colleges. Licensed practical nursing programs were also available at the vocational schools, as was training for the lower-skill health aides occupation. Almost one-half of all the graduates from the newly established community college received their degrees or certificates in health fields.

The community colleges also took over the primary responsibility for technical training during this period—promulgated by the merger of a four-year college and a four-year technical institute. A new mission of the public university, created by this merger, was to expand graduate-level education, which heretofore was available locally on a very limited basis. As resources at the university were reallocated to emphasize graduate programs, associate's degree programs were being phased out. Consequently, whereas four-year colleges accounted for 40 percent of all occupationally trained graduates in the Lowell area in 1970, by 1982 their share was less than 10 percent.

The provision of office skills was the most widely shared training responsibility in the local educational network. This institutional pattern was also changing, however, with secondary-level programs becoming increasingly more important. As in the technical field, the public university was reducing its role in the provision of office skills. The role of the public community colleges in office training was also declining, while that of the vocational schools was growing. For example, in 1970 data processing was available only at the four-year technical institute. Community colleges began offering data processing during the 1970s. By 1982, however, the vast majority (98 percent) of such training programs were provided at the vocational schools. The secretarial programs were experiencing a similar trend—with growth concentrated in the community colleges in the earlier half of the period studied, and in the vocational schools thereafter. The local high schools were also taking on a greater share of both the data-processing and secretarial education programs in the area.

Private-sector institutions in the local economy played a relatively small role in supplying occupationally trained graduates during the reindustrialization process. Moreover, their share of graduates dropped from 20 percent in 1970 to 8 percent in 1982.

This decline is attributable primarily to growing public-sector competition. As noted earlier, the private, hospital-affiliated nursing program was being supplanted by the community college program. In the Lowell area, as throughout the country, private, hospital-affiliated diploma schools of nursing have found it increasingly difficult to compete with nursing programs offered at significantly lower tuition rates to students in public institutions. In addition, the degree and scope of change in public institutions providing occupational education in the area during this period—including the opening of the new community college and regional vocational schools, and the merger of the state college and technical institute—most likely hindered the entry of private institutions.

The relative lack of private institutions in the area may also have been due to the unsettled nature and limited quantity of demands for skills in certain fields. Private, for-profit educational institutions, in particular, have been cited in studies as leaders in providing newly emerging skills.[9] Demand, however, must be sufficiently large to sustain tuition-dependent investments in plant and equipment. As noted in Chapter 5, employers hired graduates from private technical institutes outside the Lowell area prior to the establishment of local programs. These demands were generally for specialized professional or technical skills workers, and involved a small number of workers.

Three private educational institutions—extension campuses of two junior colleges, plus a proprietary school offering training in word processing—did open just at the close of this study. One of the new junior colleges offering business and office training (including computer skills) is leasing its equipment from one of the regional vocational schools.

ISSUES OF EQUITY

Consistent with the relatively large influx of female workers into the Lowell labor market in the past two decades was a growing proportion of women in the occupational education network. By 1982, women accounted for over half of the occupationally trained graduates in the local area, ten percentage points higher than in 1970.[10] In terms of integration, by the end of the period studied, women accounted for the majority of the graduates in the traditionally male management, painting and decorating, upholstery, and radio and television production programs. In the technical fields, women constituted 20 percent of the graduates by 1982, compared to less than 3 percent in 1970. In the rapidly growing computer technology area, women represented almost one-half of the graduates in 1982.

At the conclusion of the study, however, health, office, and cosmetology programs still accounted for three-quarters of all occupationally trained female graduates. Technically trained graduates accounted for less than 3 percent of all female graduates. In addition, exclusive of the cosmetology program, women represented less than 10 percent of the graduates from the trade and industry programs. Moreover, training in 45 percent of the occupational fields remained sexually segregated in 1982 — nine fields had no female graduates and thirteen fields had no male graduates.

Women comprised a disproportionately small share of the high-technology program graduates. Approximately one-quarter of the graduates from these programs between 1970 and 1982 were female; 85 percent of them received training in the office programs.[11] In the high-technology technical and trade and industry programs, women accounted for 12.0 and 3.1 percent, respectively, of the occupationally trained graduates.

The majority of the programs in this occupational education network offered training for occupations that had above-average earnings in the local labor market. Women, however, were concentrated in those programs providing skills for relatively low-paid occupations. In 1980, for example, women accounted for the bulk of all graduates in fifteen of the sixteen training programs related to occupations with below-average earnings in the local labor market.

Interviews with local educators suggest that the relative lack of women in occupational training programs geared toward the higher-paying jobs, particularly those in the trade and industry field, is the result of (1) a small number of women applying for such training and (2) the dropping out of women who did enroll. In this study, there was often only one woman enrolled in the "nontraditional" trade and industry programs; in cases where there were more, they generally lived in different school districts and took different buses to the regional vocational schools. As a result, a "critical mass" or "support group" for these young women was missing. The importance of this type of support system was emphasized by educators involved with programs that had been successfully integrated, such as the painting and decorating and graphic arts programs.

SUMMARY

The reindustrialization of the Lowell area demonstrates how educational institutions can respond to changing labor market needs and technological change in the local economy. During the period of economic renewal, the local occupational education network expanded rapidly and evolved in ways highly reflective of current labor market trends, both in general and with respect to the skill needs of the local high-technology industries. Educational institutions monitored local employment needs in Lowell through close employer-school interaction and ongoing analysis of job placement rates and student enrollment by program. The resulting patterns of occupationally trained graduates were more attuned to current occupational needs than had curriculum change been based solely on past employment trends or on occupational projections.

The Lowell study provides additional empirical support for skill-training life cycles as technologies evolve. It does so first in the transfer of skill training from the workplace to the formal educational network, and second in the changing responsibilities of various institutional components within that network as skills become more standardized and in greater demand.

APPENDIX 6A:
THE RESPONSE OF HIGHER EDUCATION TO EMPLOYMENT TRENDS IN THE LOWELL LABOR MARKET AREA

Degrees Awarded

Approximately 17,000 four-year undergraduate and graduate degrees were awarded in the Lowell labor market area (LMA) between 1970 and 1982. (See Table 6A–1 and Table 6A–2.) Bachelor's degrees accounted for the vast majority (88.3 percent) of these degrees. Master's degrees comprised 11.5 percent, and doctorates represented less than 1 percent of the total.

Bachelor's Degrees

The number of bachelor's degrees awarded in the LMA between 1970 and 1982 rose on an average 5.6 percent per year. Growth was substantially faster in the earlier half of the period (+8.7 percent) than in the latter half (+1.7 percent). By discipline, the rates as well as direction of change varied considerably. Degrees in health, management, and liberal arts disciplines grew well above average, at rates of +140.7 percent, +21.5 percent, and +14.7 percent per year, respectively. As a result, the proportion of health graduates of all four-year degrees awarded rose from barely over 1 percent in 1970 to almost 12 percent in 1982; management more than doubled its share from 12.6 percent to 26.8 percent; and the proportion of liberal arts graduates grew from 12.2 percent to 20.1 percent. (See Figure 6A–1.)

These health degree figures are somewhat misleading in terms of the supply of newly trained graduates. Nursing degrees, which accounted for two-thirds of the degrees in the health field, were in many instances awarded to experienced nurses who had returned to college for a relatively brief period to upgrade their educational training to a four-year degree. This trend is indicative of the movement nationwide for registered nurses (RNs) to complete a four-year college degree program instead of obtaining the more traditional three-year diploma from a hospital-affiliated school of nursing.

In the management field, business administration programs experienced the most rapid growth (+ 42.5 percent) throughout the period. In addition, new programs were introduced in accounting, banking, and economics. In the liberal arts, nine new programs were introduced to complement the offerings in English, history, modern languages, and music.

In contrast to the health, management, and liberal arts disciplines, growth in four-year degrees awarded in the engineering and pure and applied science disciplines was modest – slower than three percent a year from 1970 to 1982. By 1982 engineering still accounted for the largest share of bachelor's degrees awarded; however, its relative share had fallen from 39.4 percent in 1970 to 31.7 percent. Within the engineering field, however, experience was mixed. Two programs, paper engineering and textile engineering, were phased out; one new program, industrial engineering, was introduced. In addition, several established engineering programs experienced a sharp reversal in graduates within the period: the number of civil engineering graduates grew 25.9 percent a year from 1970 to 1976, but declined by almost 10 percent a year between 1976 and 1982. The nuclear engineering program experienced a similar pattern. In contrast, degrees in electrical engineering and plastics engineering declined from 1970 to 1976 but increased thereafter.

The number of pure and applied science degrees awarded declined relatively throughout the period, and in absolute terms as well after 1976. Specifically, between 1976 and 1982, the number of mathematics degrees awarded fell 6.8 percent on average per year, physics declined by 11.5 percent a year, biology declined 11.7 percent a year, and meteorology fell 9.7 percent a year. The recent decline in the number of pure and applied science graduates was mitigated by the introduction of a computer science program and an information systems program. By 1982, however, these two programs accounted for less than one in five pure and applied science graduates. Finally, the number of education degrees awarded fell both absolutely and

Table 6A–1   Bachelor's Degrees Awarded, by Discipline in the Lowell Labor Market Area, 1970–1982.[a]

Discipline	1970	1971	1972	1973	1974	1975	1976	1977	1978	1979	1980	1981	1982	Total
Engineering/tech., total	320	305	349	347	359	362	351	439	385	409	360	376	432	4,794
Chemical eng.	32	41	20	22	31	36	32	29	29	43	30	28	43	416
Civil eng./tech.	38	66	67	83	82	87	97	65	65	39	54	33	39	815
Electrical eng.	93	70	94	96	88	77	80	112	85	88	82	95	89	1,149
Industrial tech.						29	30	49	41	47	32	40	43	311
Mechanical eng.	57	37	49	71	67	62	58	64	69	77	63	73	90	837
Nuclear eng.	21	19	37	42	29	28	24	35	22	25	14	14	8	318
Paper eng.	7	8	7	2	17	—	—	—	—	—	—	—	4	45
Textile eng./tech.	21	14	4	3	5	1	—	—	—	—	—	—	—	48
Engineering, not specified								47	47	61	45	44	61	305
Health professions, total	9	0	42	45	56	61	58	66	71	115	118	146	161	948
Health education	9	0						0	0	9	19	23	20	80
Medical tech.					4	6	3	5	11	17	18	17	23	104
Nursing			42	45	52	55	55	61	60	73	67	65	65	640
Physical therapy												27	30	57
Health service admin.										16	14	14	23	67
Education	196	185	193	1	23	33	121	97	84	76	61	30	15	1,115
Management, total	102	135	148	183	214	243	216	234	282	340	336	340	365	3,138
Accounting						59	52	80	99	136	104	85	88	703
Banking									2	1	0	2	2	7
Business admin.	41	73	102	147	176	142	113	138	156	188	217	235	250	1,978
Economics						9	17	9	11	5	6	6	13	76

														Total
Industrial mgt.	61	62	46	36	38	33	34	7	14	10	9	12	12	374
Pure and applied sciences, total	86	90	70	131	181	170	175	184	152	150	132	129	115	1,765
Biology	8	9	11	15	42	31	44	42	33	37	25	33	13	343
Chemistry	24	17	14	21	15	27	17	20	16	27	15	6	22	241
Computer science													2	2
Environmental sci.					2	5	10	15	15	13	19	9	10	98
Information systems									11	4	12	14	18	59
Mathematics	27	34	27	72	78	67	54	50	37	40	37	35	32	590
Meteorology	9	12	6	6	7	17	24	13	15	13	10	11	10	156
Physics	18	18	8	8	16	7	13	37	13	14	6	13	4	176
Radiological health sci.			4	9	18	16	13	7	12	2	8	8	3	100
Liberal arts, total	99	137	67	228	260	270	212	286	286	297	308	320	274	3,044
American studies						1	10	9	9	8	5	4	4	50
Art				15	6	20	9	21	14	26	22	23	10	166
Criminal justice										10	29	58	49	146
English	27	37	25	46	45	33	19	26	20	32	16	27	39	392
History	18	31	28	36	27	30	9	7	13	22	26	31	27	305
Modern languages	1	8	3	9	17	10	6	9	9	12	23	11	14	132
Music	53	61	11	37	57	92	57	78	95	89	83	87	56	856
Philosophy					1	5	4	4	2	2	2	3	1	24
Political science				2	4	6	11	14	10	2	—	—	—	49
Psychology				48	74	36	48	35	54	58	50	38	50	491
Public service								29	20	0	18	12	3	82
Sociology				35	29	37	39	54	40	36	33	26	17	346
Other, liberal arts											1	0	4	5
Total degrees/awarded	812	852	869	935	1,093	1,139	1,133	1,306	1,260	1,387	1,315	1,341	1,362	14,804

a. 0 indicates that no one graduated from the program; — indicates that the program no longer existed.

Table 6A–2   Graduate Degrees Awarded, by Discipline in the Lowell Labor Market Area, 1970–1982.[a]

Discipline	Year													Total
	1970	1971	1972	1973	1974	1975	1976	1977	1978	1979	1980	1981	1982	
Master's Degree programs, total	40	53	53	106	90	116	203	206	188	194	218	224	252	1,943
Engineering, total	26	36	35	77	49	56	62	98	44	49	68	97	96	793
Chemical eng.								3	2	6	3	8	7	29
Computer eng.							7	11	8	18	26	34	31	135
Electrical eng.	9	15	13	16	11	13	14	25	12	8	10	10	8	164
Environmental studies					2	3	4	14	3	3	7	4	7	47
Mechanical eng.		2	2	13	4	0	5	6	2	1	3	10	9	57
Nuclear eng.			2	1	4	2	7	10	5	0	1	5	4	41
Paper eng.	1	0	1	0	1	1	1	2	0	0	1	2	2	12
Plastics eng.	3	3	5	28	15	24	16	23	7	9	12	18	17	180
Software eng.													5	5
Systems eng.			1	7	2	3	1	0	0	1	3	5	4	27
Textile eng./tech.	6	6	6	4	6	3	1	0	0	1	1	0	0	34
Health professions														
Gerontology								17	21	16	16	18	20	108
Education, total					10	22	103	60	81	91	77	49	44	537
Education							74	34	52	69	64	39	37	369
Mathematics for Teachers					10	22	29	26	29	22	13	10	7	168
Management														
Bus. admin./mgt. science			3	3	14	9	5	12	11	13	20	19	34	143

	14	17	15	26	17	29	23	18	20	18	23	27	37	
Pure and applied sciences, total	14	17	15	26	17	29	23	18	20	18	23	27	37	284
Biology				4	4	6	1	3	5	5	5	8	5	32
Chemistry		5	6	4	5	11	3	6	4	7	4	4	4	57
Computer science				9	5	6	0	0	0	0	0	6	16	42
Mathematics		2	4	7	2	3	6	3	6	0	1	1	2	45
Physics		5	4	0	1	2	6	1	3	3	3	1	2	40
Polymer science		5	1	0	0	1	4	0	2	3	0	2	1	21
Radiological health sci.				2	0	1	3	5	0	0	10	5	7	33
Other[b]	14													14
Liberal arts														
Music							10	7	11	1	14	14	21	78
Doctorate programs, total	3	0	0	0	6	2	8	1	2	3	4	3	6	38
Chemistry	2	0	0	0	1	0	3	0	1	2	1	2	3	15
Physics	1	0	0	0	3	1	5	1	1	1	3	1	3	20
Polymer science		0	0	0	2	1	0	0	0	0	0	0	0	3

a. 0 indicates that no one graduated from the program; — indicates that the program no longer existed.
b. Distribution not available.

Figure 6A–1 Distribution of Bachelor's and Master's Degrees in the Lowell Labor Market Area, by Discipline, 1970 and 1982.

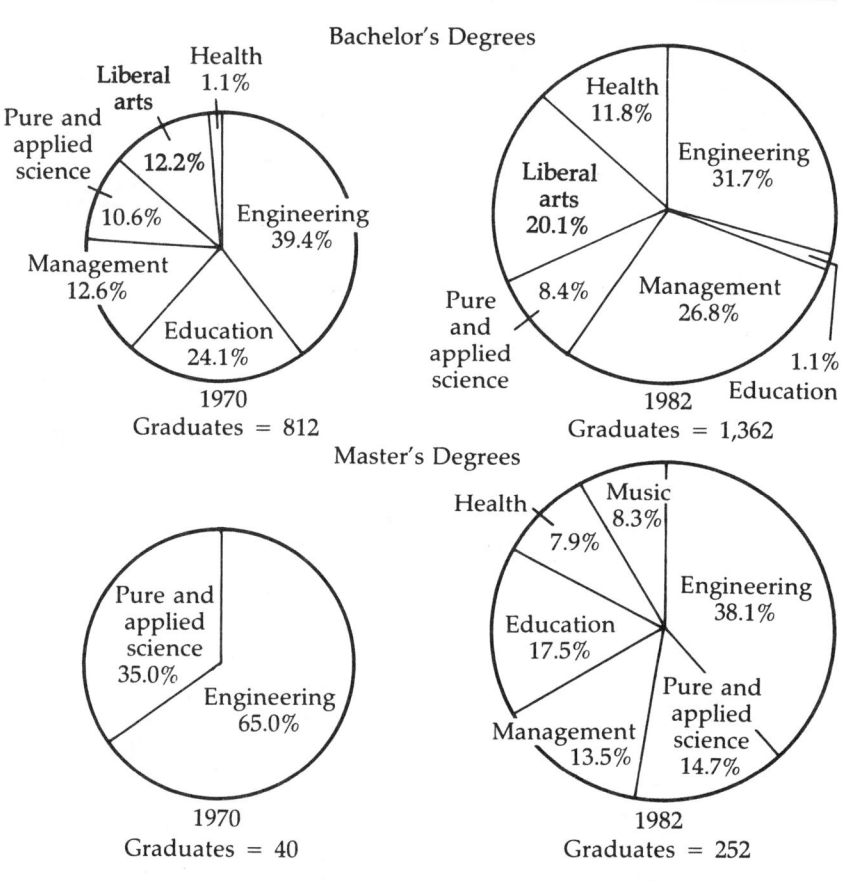

relatively: they declined 7.7 percent a year on average from 1970 to 1982 and decreased in share from almost one-quarter of all bachelors degrees to less than 2 percent.

Graduate Degrees

The number of master's degrees awarded grew considerably more rapidly (+44.2 percent a year) than bachelor's degrees awarded from 1970 to 1982. At the beginning of the period, master's degrees were offered only in engineering and in pure and applied science. These two fields, however, grew relatively slowly during the period, and by 1982 they accounted for only one-half of the masters degrees conferred. New programs in education, management, music, and health were responsible for the growth at this level, accounting for 17.5 percent, 13.5 percent, 8.3 percent, and 7.9 percent, respectively, of the master's degrees awarded in 1982.

Relatively few doctorates were awarded in the local area between 1970 and 1982. Of the 38 total Ph.D.'s, 20 were awarded in physics, 15 in chemistry, and 3 in polymer science. There was no discernible trend in doctorate degrees over the period.

Employment Trends in Professional Labor Markets

Unlike the majority of occupational education graduates below the four-year college degree level, recipients of bachelor's and graduate degrees often face labor markets that are statewide, regional, or national in scope. In addition, specific occupations "related" to particular degree programs, such as business administration or those in the pure and applied sciences, are much less clearly defined than those matched with occupational education programs below the four-year degree level. Thus instead of replicating the analyses on educational responsiveness to local labor market trends conducted in Chapter 3, this section briefly discusses the changes in the supply of bachelor and graduate degrees awarded in the context of more general employment trends for these college-educated workers.[12]

Employment in the health field, the fastest-growing area of college graduates, was strong in the 1960s. Employment growth in this field locally as well as nationally was projected to be well above average. Although health employment grew rapidly nationally in the 1970s, however, local growth was relatively slow. As noted previously, though, these data on health graduates overstate the supply of newly trained health practitioners, as many of the graduates were already nurses employed in the Lowell area.

As in the health field, the management field experienced relatively fast growth of both college graduates and jobs in the 1960s, and employment was projected to grow at above-average rates into the 1990s. In contrast to the health field, the number of jobs in management grew relatively rapidly during the 1970s in the local economy as well as nationally. Estimating potential surpluses or shortages of management graduates, however, is difficult even on a national level, since many recruiters regard a wide range of college majors, including liberal arts, as preparation for managerial positions.

There is debate over whether the market for MBAs will soon reach a saturation point.[13] Given the continuing increase of MBAs and their relatively high salaries compared to bachelor's degree candidates in business and related fields, employers could shift more of their demands toward the less expensive bachelor's degree holders. The trend toward rising educational requirements in general, however, suggests a more plausible scenario: that employers will continue to expand their demand for MBAs but that the wage differential between the graduate and undergraduate degrees in business will shrink.

In the accounting field, the job market was favorable throughout the 1960s and 1970s and was projected to be so through the mid-1990s. If, as appears to be the case, recruiters are now seeking accounting majors with four-year college degrees instead of the more traditional two-year degree, the market for these graduates will be even stronger than that projected.

In the engineering field, labor market conditions changed considerably from 1960 to 1980. In local and national labor markets the employment of engineers grew relatively fast in the 1960s but well below average in the 1970s. In the late 1960s and early 1970s Massachusetts, particularly along Route 128, was especially hard hit by defense budget cutbacks, which resulted in widespread layoffs of engineers. In contrast to the 1960s, relative earnings of engineers declined during the early 1970s. When the demand for engineers rebounded in the mid-1970s, the number of graduates was falling and widespread shortages were reported.[14] The relatively slow growth of graduates in engineering in the Lowell labor market area in the early 1970s is reflective of labor market cycles in this field. Since 1976 the number of engineering graduates in the labor market area at both the bachelor's and master's degree levels has risen at rates well above average.

Graduates with pure and applied science degrees, another slow-growing field between 1970 and 1982, seek a wide range of professional positions. A currently popular view is that the number of graduates in this field falls far short of the science and technology needs of the country.[15] Shortages of mathematics and science teachers in particular have been reported in over forty states.[16]

Education graduates, more generally, entered favorable job markets in the 1960s, but their counterparts in the 1970s were not so lucky. The depressed labor market for teachers, brought on primarily by the declining school-age population, was felt in elementary schools in the early 1970s and then in the secondary schools in the late 1970s. Although the population is again increasing as the post-World War II "baby boom" cohort has reached childbearing age, employment prospects for teachers in fields other than mathematics and science were expected to remain weak through the mid-1980s.

Although the total number of degrees awarded in education in the Lowell area continued to decline between 1970 and 1982, the proportion of education degrees at the graduate level rose. Labor market conditions for teachers in the 1970s has made graduate school more appealing for education majors. In addition, graduates with master's degrees are in a better competitive position for the relatively few jobs now available. Furthermore, given the change in the composition of the available relevant labor supply, when the market for teachers picks up again in the late 1980s and 1990s, schools may favor teachers with graduate degrees for entry-level positions.

In sum, recent trends in bachelor's and graduate degrees awarded overall in the Lowell labor market area have been responsive to employment trends in these fields. Skill imbalances result primarily from shifts in employment demand, changes in relative earnings of occupations, and training lags. Unfortunately, placement data are not available for these college graduates. Without knowing whether these graduates are getting jobs—and if so, where and in what occupations—it is difficult to address issues regarding expansion or contraction of particular programs in the Lowell area to meet changing local and regional employment needs.

APPENDIX 6B:
MODEL OF EDUCATIONAL RESPONSE

A grid (Figure 6B–1) relating employment and occupationally trained graduates facilitates the analysis of educational response to labor market trends. The graduates during the period studied (1970–1982) are divided, by type of occupational training, into four groups: (1) programs that had declined absolutely as well as relatively in terms of graduates or had been eliminated during the period; (2) programs in which the percentage of change in graduates was positive but below average; (3) programs in which the percentage change in graduates was average; and (4) programs in which the percentage change in graduates grew more quickly than average, or programs that had been introduced during the period and offered a type of training previously unavailable in the area. Changes in employment by occupation (1970–1980) are also divided into four categories: (1) occupations that had declined in the local economy; (2) occupations that had below-average growth; (3) occupations that had average growth; and (4) occupations that had above-average growth.

If changes in the occupational education network were perfectly "on target" with those in the local labor market, one would find the various programs positioned in the four cells along the diagonal from the upper left corner to the lower right corner—indicated in black on the grid. The first cell, for instance, would list occupations that had declined in employment and for which the number of graduates had fallen. The blackened diagonal alignment would indicate that occupational training was no longer available or that the number of trained graduates was shrinking for occupations whose numbers were declining locally, and that the new and rapidly growing programs were occupations for which employment was growing relatively rapidly.

Appendix 6B is reprinted with permission from Patricia Flynn, "Technological Change, the 'Training Cycle,' and Economic Development," from John Rees (ed.) *Technology, Regions and Policy* (Totowa, New Jersey: Rowman & Littlefield, 1986), pp. 303–306.

Figure 6B–1   Occupational Education and Labor Market Trends

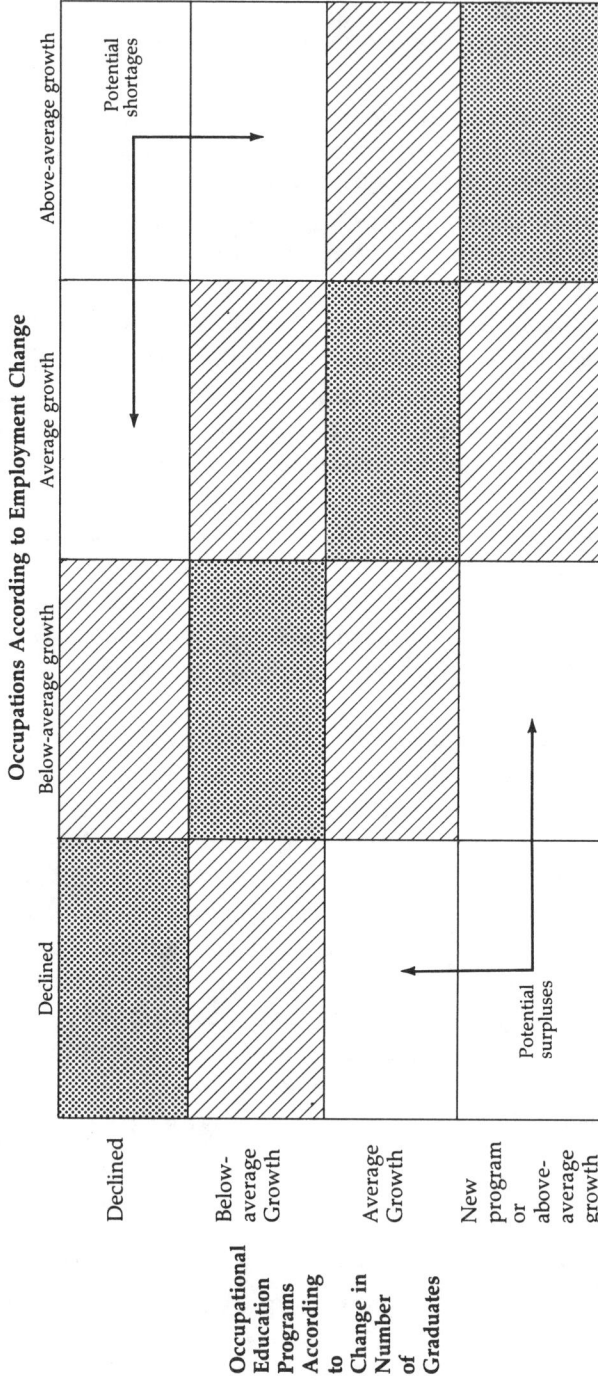

Source: Patricia Flynn, "Technological Change, the 'Training Cycle,' and Economic Development," from John Rees (ed.) *Technology, Regions and Policy* (Totowa, New Jersey: Rowman & Littlefield, 1986). p. 304. (reprinted with permission).

Several of the other cells along the "fringe" of this diagonal —
shown by the striped cells in the grid — suggest "reasonable"
responses of the occupational education system to the labor
market trends. For instance, programs with declining numbers
of graduates in areas where employment has shown only modest growth is reflective of the labor market change. Furthermore,
the introduction of new fields of training in an area of employment experiencing average growth indicates that the educational institutions are beginning to address employer needs that
the local educational system has not been meeting. In times of
rapid growth, as was the case in the Lowell area during the
study period, "average" growth can represent a substantial
number of jobs. Even in areas where employment has declined,
replacement needs continue and may warrant relatively slow
growth of the number of graduates trained in these areas.
Replacement demand is particularly important in the more
traditional, blue-collar manufacturing occupations in which the
average age of workers tends to be well above the average
for all occupations. These are jobs often requiring occupational
skill training.

The six "outlier" (white) cells in the grid signal potential
"problem" areas in the provision of skilled workers. Mismatches
in the supply and demand of occupational skills may occur
because occupational training has been eliminated or has decreased in areas of relatively fast employment growth. In such
instances, represented by the three cells in the northeast corner
of the grid, occupational skill shortages may occur, hindering
further economic growth. Alternatively if training programs
have been introduced or programs are experiencing rapid
growth in areas of relatively slow or declining employment,
potential surpluses of trained graduates may result. This latter
case is represented by the three cells in the southwest corner of
the grid.

It is critical that educators and policy makers recognize that
outliers signal *potential* problem areas. The mere presence of
outliers is not an immediate "call for action" to introduce or
expand programs in the cells designating potential shortages, or
to eliminate or contract programs in the cells signaling potential
surpluses. Whether shortages or surpluses of trained graduates

actually occur in the local labor market in these occupations depends on a variety of factors, including skills available from other sources of labor supply, such as the unemployed, reentrants to the labor market, four-year college degree holders, and current employees; whether the skills needed to perform the job are best acquired on the job or in a school setting; the hiring and training practices of local employers; in-migration and out-migration of workers in various occupations; and economic conditions in the economy overall. Without analyses of these factors, statistics on the supply and demand of occupationally trained graduates may be misleading and suggest erroneous policy implications.

Figure 6B–2 indicates the results of this analysis for the Lowell area. Additional information on this model of educational response is available in Patricia M. Flynn, *Production Life Cycles and Their Implications for Education and Training*, final report, grant number NIE-G-82-0033, National Institute of Education, February 1984.

Figure 6B–2 Occupational Education and Labor Market Trends, Lowell Labor Market Area

Occupations According to Employment Change, 1970–1980

Occupational Education Programs According to Change in Number of Graduates, 1970–1982	Declined	Below-average growth	Average growth	Above-average growth
Declined	Civil tech. Industrial tech. Mechanical tech. Other tech.[a]		Electrical tech.	
Below-average growth		Auto mechanics Cosmetology Machine shop	Nursing (LPN) Carpentry	Home economics
Average growth	Upholstery		Electricity	
New program or above-average growth	Dental hygiene Drafting Medical tech. Metal Fabrication Clothing mgt. Painting and Decorating Sheetmetal	Auto body	Computer tech. Dental asst. Distrib. ed. Electronic tech. Electrical/mech. tech. Electronics Fire protection Health services Law enforcement Management Masonry Medical aide Medical records Plumbing	Accounting Child care Commercial art Data processing Food mgt. Graphic arts HVAC Mechanics, exc. auto Public services Radio & TV Secretarial Welding

Source: Patricia Flynn, "Technological Change, the 'Training Cycle,' and Economic Development," from John Rees (ed.) *Technology, Regions and Policy* (Totowa, New Jersey: Rowman & Littlefield, 1986), p. 306 (reprinted with permission)

a. Includes graduates from programs in environmental technology, applied mathematics, applied chemistry, and engineering science technology.

IV CONCLUSIONS

7 TECHNOLOGICAL CHANGE AND HUMAN RESOURCE PLANNING

Changes in skill requirements, training needs, the industrial and occupational mix of employment, and the spatial location of jobs are indicative of technological change and the dynamic character of production processes. These changes in turn affect employers' hiring and staffing patterns, workers' career paths, and economic growth and development. The ramifications of these phenomena are pervasive: they affect local as well as regional and international markets, high-technology as well as mature industrial sectors, economically depressed as well as "boomtown" economies, and high-skilled as well as low- and unskilled workers.

The impacts of technological change are both positive and negative. Technological change is a key factor in bolstering productivity, in creating jobs and advancement opportunities, and in promoting economic growth. Nevertheless, it also causes the deskilling and elimination of tasks, resulting in skill obsolescence, worker displacement, layoffs, and unemployment.

This book presents a new, highly disaggregated approach to understanding the wide-ranging and ever-changing effects of technological change on jobs and workers. Extending the traditional product, process, and technology life-cycle models to an analysis of skills, training, occupations, and jobs, it presents an analytical framework within which to explore the effects of technological change systematically. Common patterns and trends in these work-related variables as technologies evolve are identified through this conceptual mode. Empirical evidence drawn from both a large international body of case-study materials and new field research at the local level is presented in support of the framework. The findings reconcile many of the apparent inconsistencies and inconclusive findings in the aggregate literature on technological change. Moreover, the life-cycle framework offers new insights into human resource planning to facilitate technological change. This chapter presents the major conclusions and policy implications derived from the study.

TECHNOLOGICAL CHANGE AT THE WORKPLACE

When seen in the evolutionary framework of product and technology life cycles, many aspects of technological change at the workplace can be expected to have an uneven impact on skills, training, and staffing patterns. For example, a firm that adopts a new technology will experience different demands for skills and training than a firm that adopts the same technology after it has matured. The introduction of new and emerging technologies generates highly skilled scientific, engineering and production tasks. The introduction of increasingly more complex equipment as products and production processes become standardized creates more generic semiskilled technical and maintenance tasks. Combined with increasing levels of output, this standardization results in the deskilling of various tasks when relatively mature technologies are involved.

The Skill-Training Life Cycle

A "skill-training life cycle" highlights the evolving nature of training needs and the responsibility of their provision at different stages of a technology's development. The firm-specific nature of skills required and the lack of workers with these skills in the earliest phase of the skill-training life cycle, means that employers must provide their own training or rely on the equipment manufacturer to do so. In the second and third phases of this cycle, technologies become diffused and products and equipment become standardized. Firm-specific skills become general skills that are transferable among employers and the supply of appropriately trained, experienced workers increases. In addition, the growth in demand for more standardized skills encourages their formalization and "production" in educational institutions. Hence, in contrast to their early-adopting counterparts, employers who adopt relatively mature technologies can hire trained workers—new graduates as well as experienced workers—from outside the firm.

In the final phase of the skill-training life cycle, as old technologies give way to new, training focuses on replacement needs and on the retraining of workers for other fields. A limited market for skills and student enrollments may result in the termination of training programs in these fields—shifting the focus of training back to the firm.

Staffing Patterns

Changes in skill and training needs influence firms' staffing patterns—which in turn affect workers' career paths. In particular, the likelihood that the adopting firm's employees will be selected, trained, and upgraded to perform the more high-skill tasks created by technological change will decline as the technology matures. Uncertainty about the nature and quantity of skilled workers they will need for newly emerging tasks leads early adopters of technologies to incorporate such tasks into existing jobs rather than disrupt their current job structures. Empirical evidence shows that workers whose jobs have been "enlarged" are likely to continue to staff these positions. If new positions must be created by these early adopters, as is often the case, for instance, when skill requirements change considerably, employers generally prefer to staff the new high-skill jobs through internal transfer or promotion rather than hire un-trained, unknown workers.

In contrast, the skill needs generated by the adoption of relatively mature technologies are likely to be associated with well-defined occupations that have specific educational and related work experience criteria for entry. In such cases, the staffing of the new positions may well result in current employees being passed over for trained workers from outside the firm. More generally, with the establishment of new entry ports and formal educational requirements, the adoptions of relatively mature technologies can create institutional barriers for upgrading and promotion within the firm.

TECHNOLOGICAL CHANGE IN THE COMMUNITY

The production life-cycle models suggest that a variety of local employment patterns exist at any one time. Regions and local economies differ considerably in terms of resource distributions and relative cost structures, and hence in their attractiveness to employers for locating various production activities. In addition, the mix of technologies and products in the employment base of geographic areas will change over time. Employers relocate production activities in response to changing resource requirements and a growing cost-sensitive competitiveness as technologies mature. New jobs replace more traditional employment opportunities. The life-cycle framework suggests that the key to understanding the likely impacts of technological change in a community is analysis of the mix and characteristics of firms and human resources in the particular area.

High Technology

The life-cycle models provide a new perspective from which to assess the recent focus on high technology as a key to local economic development. The Lowell area's reindustrialization supports the view that high-technology industries can provide the engine to economic growth and development.

Industries in general encompass a variety of technologies and products spanning a range of phases of development and requiring different skills and training needs. As usually defined, "high technology" includes a list of industries selected on the basis of their relatively high proportions of R&D expenditures and of professional and technical workers. Blue-collar and clerical jobs, however, account for the majority of the employment in these newer industries. Moreover, there is considerable diversity both among and within high-technology industries with respect to occupations, wage rates, firm sizes, and organizational ownership patterns.

From a life-cycle perspective, the concept of a high-technology industry is misleading. "High technology" is a dynamic and relative concept that describes the early phase of industrial development. "High-technology employment," viewed in this framework, refers only to jobs involved with R&D, innovation, and nonstandardized production activities. Industries, or components thereof, pass through high-technology phases — characterized by rapid technological change, a relatively high degree of R&D expenditures, and a dependence on highly skilled workers.[1] Whereas the textile industry is often referred to as mature or traditional, it represented a "high-technology" industry a hundred years ago. Similarly, industries considered "high-technology" today, such as computer, powdered metals, biotechnology, or information processing, may or may not be the high-technology industries of tomorrow.

Employment Diversity

Production life-cycle models accentuate the fact that technological "progress" can be detrimental to the economic growth and development of individual communities. The vulnerability of a particular community to the destabilizing effects of technological change — such as industrial decline, job relocations, and plant closings — depends on the area's mix of businesses and industries. A diversified employment base provides alternative local employment opportunities to counteract job loss due to the deskilling process or production shifts to other locations.[2] In contrast, an area whose employment is tied to one or two product lines or to a group of firms with products and technologies at similar stages of development may experience significant swings in economic activity as technologies mature.

Regional specialization within firms and industries has generated a relatively high proportion of R&D, design, and small-run production activities in the Lowell labor market, highlighted in Chapters 5 and 6, compared to other areas. One industry—the office, accounting, and computing machines industry—however, has dominated the reindustrialization process in Lowell. Moreover, one employer, Wang Laboratories, accounted for approximately one-half of the local employment in this industry in the early 1980s.

Labor and skill shortages attributable to the growth of Lowell's newer industries have spilled over into more traditional employment sectors, hastening the departure from the area of some of these jobs. Moreover, in the last two years layoffs attributed at least in part to slumping demand and increased competition among computer firms have occurred locally.[3]

The combination of production life cycles and the potential crowding-out effects of successful new employment accentuates the importance of a diversified employment base for building a foundation for long-term economic growth in the community.

POLICY IMPLICATIONS FOR HUMAN RESOURCE PLANNING

The production life-cycle framework accentuates the need for planning human resource policies for both the "up side" and the "down side" of technological change. Failure to adapt to the newly created skill needs generated by technological change can restrict the productivity of workers and of firms, undermining industrial competitiveness and economic growth. Failure to minimize the negative impacts of such change, as jobs are simplified and eliminated, can further constrain the benefits of technological progress.

The findings of this study suggest a series of policy implications for employers, educators, and economic planners seeking to facilitate technological change.

Technological Change at the Firm

Trends in skill requirements, training needs, and occupations as technologies and products develop affect the larger environment in which firm-level decisions concerning human resource adjustments are made. Managers can facilitate technological change through planning that integrates the "natural" shifts in skills with training needs.

Management practices and labor-management agreements continue to play key roles in determining the effects of technological change at a particular worksite.[4] The timing of the adoption, as well as ways in which work tasks are allocated among jobs and workers, for instance, affects the nature of new skill requirements and the degree of difficulty in reallocating workers displaced by the change.

Timing

One key element in the timing of the technological change at the workplace is the state of the local economy. An expanding employment base provides more alternative job opportunities for displaced workers. Economic growth will facilitate voluntary quits, thereby reducing the pool of workers seeking reassignment within the firm.

The level of product demand at the time of technological change affects the firm's ability to reallocate workers within the firm. Many employers plan technological changes to coincide with business expansions, permitting greater opportunities for promotions and transfers within the firm. In contrast, stable or declining sales diminish the firm's internal options for reassigning displaced workers. When technological change involves the geographic relocation of jobs, empirical evidence shows that employees at all skill levels will often decline transfers that require a residential move or a long daily commute.

The timing of the adoption relative to the technology's life cycle further determines the need for human resource adjustments at the firm. As discussed earlier, the stage of development of the technology influences not only the nature of new skill needs, but also the availability of appropriately trained workers.

Managers may choose to postpone adopting a technology until they are more certain about the quantity and quality of new skills required. But as skill needs become clearer, the likelihood of disrupting internal job ladders increases. The disruption or curtailment of advancement opportunities within the firm can increase costly worker turnover.

Distribution and Allocation of Work

While technological change directly affects the level of skills required to perform various tasks, the manner in which these changes are integrated into jobs reflects management practices and labor-management agreements at the workplace.

The deskilling of tasks need not result in the deskilling of jobs nor in the downgrading of workers. Tasks can often be regrouped to generate jobs requiring similar or more advanced skills than prior to the change. Moreover, the deskilling of a job might even lead to promotion of a worker farther down the job hierarchy.

Higher-skill jobs created by technological change also offer a range of options at the workplace. Empirical evidence suggests that the number of such jobs is likely to be relatively small. Promotion from within, however, generates additional opportunities for advancement of workers at lower levels in the job hierarchy as they staff the newly-vacated positions. Many more workers can benefit through upgrading and promotions when new positions are staffed from inside rather than from outside the firm.

Firms with relatively high rates of employee turnover have more leverage in human resource adjustments to technological change: the pool of workers seeking reassignment is reduced in size, and simultaneously more job openings are created. Case studies indicate that employers can rely on relatively high attrition, in conjunction with hiring freezes prior to the adoptions, to achieve a decline in employment without "displacement" of incumbent workers from the firm.

High rates of turnover alone do not eliminate the need for human resource planning. Management needs to be sensitive to the subset of workers whose jobs are most directly affected by the technological change. For instance, whereas relatively few layoffs were reported in the case studies on office automation, displaced workers were less likely than their counterparts in factories to be promoted to the new, better jobs. Workers subject to the negative impacts of the change with little hope of participating in the benefits are not likely to welcome or support technological changes at the worksite.

Technological Change and Education

Technological change disrupts past employment patterns and creates new labor market demands that change as technologies develop. Education and training programs can facilitate structural change by adapting to employers' diverse and evolving skill needs. Schools and colleges can help employers meet highly skilled requirements, and they can also ease the problems of technological displacement.

New and Emerging Skills

Quantitatively, the demands for new, highly skilled labor created by the adoption of new technologies appear relatively small compared to total employment needs. Failure to meet these new requirements, however, can hamper productivity gains and economic growth.

Anticipating new and emerging skill requirements is particularly difficult: they do not appear in past employment trends and are not identified by traditional forecasting techniques.[5] On-the-job training and employer-sponsored training programs are critical for the determination and acquisition of skills required in emerging fields. Working in an environment of considerable uncertainty and relatively high risk, scientists and engineers determine new skill requirements at the workplace on a trial-and-error basis. Moreover, in this early phase of development, there is a bias toward skills that are higher than would be required after initial product development.

Schools cannot hope to prepare workers for emerging skill needs as they initially arise. As skill-training life cycles evolve, however, and skills become more generalized and transferable among employers, schools can provide such training. This skill-transfer process requires closer collaboration between employers and other education and training institutions than has occurred in the past. It also highlights the need for assessing the "transferability" of various skills among workplaces.

Venture capital for small-scale experimental programs facilitates the transfer of new and emerging skills training to the schools. Strong program monitoring and evaluation, and extensive dissemination of the results, will maximize the spillover benefits from the initial venture capital. The federal government is a logical candidate to sponsor such experimental programs, in that it can coordinate them nationally, minimize duplication, and provide for widespread dissemination of the results. If federal assistance is not forthcoming, states concerned with training for new and emerging skills should take the initiative in funding experimental programs or in developing mechanisms that encourage a flow of private-sector funds for this purpose.

Preparing for Change

Education and training policies designed to facilitate meeting the needs of changing skill requirements should guard against being so "labor-market responsive" as to undermine the workers' ability to adjust to structural changes over time. Many states and regions are seeking to attract firms—particularly in high-technology fields—by promising "tailor made," or custom-designed, workforces.[6] Public programs offering employers a work force to meet their relatively specific production needs, however, tend to reduce worker flexibility in the labor market. Programs geared to youths, in particular, need to be broad enough to prepare them to work in a variety of situations.

The time frame in which planning and evaluation decisions are made often differs for educators and employers, with the latter generally shorter than the former. Educational planning for technological change needs to be sensitive to short-run and long-term labor market conditions.[7]

Generating future skill surpluses. Under the immediate pressure to fill job vacancies, it is tempting to implement quick and ambitious programs to expand the supply of trained workers, rather than relying on employers to solve some of their immediate staffing difficulties through changes in recruitment and internal training practices.

Schools and colleges can generate skill surpluses if they do not accurately predict patterns of deskilling and of job relocation patterns due to technological change. A misreading of job entry requirements can also result in trained graduates that employers view as inappropriate for their needs.

There are recent indications that many educational institutions have jumped on the high-tech bandwagon and may soon find themselves contributing to various skill surpluses in positions such as computer programmers and electronics technicians.[8] Ample evidence elsewhere suggests caution against rapid installment of programs to build up skill supplies unless the shortage is large and continued demand can be demonstrated.[9]

A local economy undergoing rapid change is particularly predisposed to structural shifts in labor demands. Educators need to supplement traditional sources of labor market data with other information, such as that from local employers and placement officials, to monitor current employment conditions. Human resource policies should address not only the existence of skill scarcities but also such issues as how rapidly shortages can be addressed and at what risk of eventually stimulating skill surpluses.

Skill Obsolescence.   Jobs and workers at all skill levels are vulnerable to the deskilling process and to technology-induced obsolescence. The vast majority of workers likely to be affected by technological change and industrial relocation are already at work—highlighting the need for retraining programs.

Empirical evidence suggests that many workers affected by the deskilling process are transferred to other positions within the firm. However, workers who are displaced *from the firm* are often in critical need of retraining and job-search assistance. Technological changes that result in plant closings and relocations of worksites, in particular, impede the process whereby workers are retrained by employers for new technologies. In addition, older workers are particularly vulnerable to the negative impacts of technological change at the workplace.

The likelihood of skill obsolescence and worker displacement, even in prosperous times, suggests the need for an ongoing local retraining capacity. Retraining programs in an area should be designed to evolve with production life cycles, recognizing innovations and developments in technologies and products as signals of future skill needs.

The Skill-Training Life Cycle and Institutions

The institutional components of the education and training network emphasize different goals, face diverse constraints, and play disparate roles in preparing workers for employment. Such diversity generates a range of institutional patterns and responsibilities and a mix of employer-school linkages within the local labor market. Moreover, as skill levels and demands change over the course of industrial development, different types of institutions are best suited to providing the training needed.

The survival of proprietary schools, for instance, depends on their ability to identify new educational markets and to respond quickly to those emerging demands. These private, for-profit schools offer a "no-frills" approach to education with few, if any, ancillary services available to students. Community colleges also provide job skills to students, but, in contrast to private schools, they offer highly subsidized training and a range of cocurricular activities to all interested students. Vocational schools provide students a cluster of job skills combined with related theoretical foundations in the classroom. In addition, vocational students usually receive on-the-job training through cooperative education programs. In contrast, much of the training provided by employers gives workers more narrow training tailored to the firm's specific employment needs.

The occupational education network, as extensive as it is, is a much smaller piece of the total training system than is generally acknowledged by public education and training policy. Public policy may be expecting the schools to do too much in terms of providing job skills while other providers of training — including firms, union apprenticeship programs, the military, and government training programs—are not given enough attention.[10]

Educational planning and policies for the provision of job-related skills should incorporate an understanding of the dynamics of the skill-training life cycle. Occupational educators working with employers and these other skill providers can foster the institutional and curricular mix that best meets the evolving needs of the community. This inclusive approach should build on the competitive advantages of the institutions involved.

Local Economic Development

Production life-cycle models suggest that education and training policy should be a cornerstone of a more broad-based economic development strategy. Economic development planning should seek to balance the needs of newer firms with those of firms in the more traditional industries—industries that account for the bulk of employment in most areas. A program that is aimed at attracting and cultivating new firms but that denies resources to more traditional employers can contribute to the "premature" departure of a significant number of jobs from the local community. The existence of spillovers—and, in particular, the "crowding-out" effect of a new, dominant high-growth industry—further highlights the need for a balanced, comprehensive human resource strategy.

High-Technology Employment

Many state and local attempts to foster a high-technology employment base focus on attracting branch plants of established companies into the area.[11] A typical approach is to entice these employers with relatively low labor costs, tax abatement schemes, or customized training programs. The production life-cycles framework and the empirical evidence suggest, however, that relatively standardized activities are much more likely than R&D and innovative activities to respond to such cost-related factors. These strategies may encourage a relatively large number of jobs to the area and hence appear immediately successful. But the relatively low- and unskilled jobs involved are particularly prone to further relocation as demands grow and competition becomes increasingly a function of production costs.[12] Research suggests that branch plants also are less likely than indigenous new firms to act as a "seedbed" or "growth pole" in stimulating spinoff firms and new employment opportunities in the area.[13]

As demonstrated in Chapter 5, the attraction of new and emerging businesses and industries can be used effectively as a development tool in economically depressed areas. High-technology employment, however, is not sufficiently large enough to "rescue" all such communities.[14] Innovation and new product development do contribute to job creation at the R&D site. Most job growth, though, is associated with increasing product demand, standardization, and lower unit production costs.[15] Moreover, recent studies suggest that many depressed communities in particular may not have the wherewithal to attract venture capital and highly skilled workers sought by high-technology employers.[16]

The key to economic renewal for many communities may lie in a different type of high-technology solution — the integration of high-technology products into more traditional industries to help them become more competitive.[17] Ongoing investment, facilitating the replacement of obsolete production processes and outdated equipment, fosters competitiveness in the established firms in an area.

Area-Specific Development Strategies

The reindustrialization of Lowell highlights the need to devise economic development strategies appropriate to the characteristics of a particular area. Many communities seeking to "reindustrialize" their economies because traditional manufacturing jobs are disappearing do not share Lowell's labor market characteristics. A study in 1982 for example illustrates the improbability that the jobs in the high-technology industries responsible for the reindustrialization of Lowell could effectively stimulate an economic turnaround in Detroit or Pittsburgh.[18]

National Competitiveness. Pittsburgh and Detroit are at a relative disadvantage nationally in competing for the types of high-technology industries that located in the Lowell area. These areas have proportionately fewer engineering and scientific workers than many other parts of the country, especially such high-technology centers as Route 128 in Massachusetts, Silicon Valley in California, and Research Triangle Park in North Carolina. More generally, Pittsburgh and Detroit lack the agglomeration economies that derive from having such a high-technology center located within commuting distance.

Detroit and Pittsburgh do have a relatively large supply of production workers, many of whom have lost jobs in the declining automobile and steel industries. Average unemployment in 1982 was 16 percent in the Detroit area and 12 percent in the Pittsburgh area. In contrast to the Lowell area, however, relative wages of production workers in these two communities are significantly above national averages. In 1982 the average hourly earnings of production workers in manufacturing were almost 40 percent above the national average in the Detroit area and 25 percent above the national average in the Pittsburgh area.[19] In some of the high-technology industries, production wages in these labor markets are more competitive. By and large, however, they remain high, particularly compared to wages in Massachusetts.

In addition, given the often-stated desire of high-technology employers to avoid unions, the relatively high proportion of organized workers in Pittsburgh and Detroit is likely to deter such employers from entering these labor markets.

Local Competitiveness. Aside from the question of national competitiveness is the issue of acceptance of newer industries within the local communities. Whereas jobs in the high-technology industries in Lowell compared favorably to alternative employment opportunities in that area, such would not be the case if these jobs located in Pittsburgh or Detroit. Production wages in the declining, traditional industries in those cities usually exceeded those paid in the high-technology firms in Lowell. In addition, while wage rates of the traditional industries in Lowell were relatively low by local standards, the auto industry in Detroit and the steel industry in Pittsburgh were still among the highest-paid sectors in these local economies in 1982. Thus workers may well incur significant cuts in pay if they were to take jobs in the newer industries, even if they were to remain in same occupation. In 1980, for example, full-time assemblers, the largest occupational group in high-technology industries in Massachusetts, were earning on average 60 percent more in Detroit ($17,145) than assemblers in Massachusetts ($10,576). Average annual earnings of assemblers in Pittsburgh ($14,090) exceeded those in Massachusetts by one-third.[20]

In addition, the industries that revitalized Lowell would offer far fewer opportunities in Detroit and Pittsburgh for workers to upgrade their occupations. Statewide data indicate, for instance, that a higher proportion of the work force in the traditional industries in Michigan and Pennsylvania are in skilled production jobs than in Massachusetts.[21] For example, over 20 percent of the workers in Michigan's auto industry and in Pennsylvania's steel industry are in skilled craft and repair occupations. This is more than double the percentage of skilled production workers in the apparel and leather industries in Massachusetts. Almost three-quarters of all workers in Massachusetts's apparel industry are in unskilled operative and laborer jobs, compared to approximately one-half of those in the auto and steel industries in Michigan and Pennsylvania, respectively.

Moreover, due to regional specialization the occupational mix of jobs in these high-technology industries differs by state, with the high-technology work force in Massachusetts more concentrated in higher-skill jobs. In the electrical and electronic equipment industry in 1980, for example, 18 percent of Massachusetts's workers were in professional and technical occupations, relative to 12 percent of Michigan's and 14 percent of Pennsylvania's. Skilled craft and repair workers also account for a higher share of the production jobs in this industry in Massachusetts than they do in Michigan and Pennsylvania.

Finally, there is the issue of timing. In Lowell, over forty years elapsed between the devastating departure of the textile industry and the resurrection of the area as a center of high-technology employment. As a result, the vast majority of the workers in the new high-technology industries were of a different generation from those who had lost their jobs in the mills. In contrast, Pittsburgh and Detroit are faced with retraining the workers who have just recently lost their jobs. Such training would be for industries vastly different from the workers' former jobs. Moreover, the unemployed auto and steel workers may find the lower-paying, relatively unskilled new jobs unacceptable. They may also find it difficult to adjust to full-time or relatively long-term training programs to acquire the skills to qualify for the better of the new jobs created.

As demonstrated with workplaces, the specific ways in which a community is affected by the common trends in skill, training, and employment over the technology life cycle depend on the characteristics of the particular area. The Lowell-Detroit-Pittsburgh comparison highlights the critical need for local labor market analysis in assessing economic development strategies. The types of jobs that will reverse economic decline in one area may or may not do so in others. What "succeeds" in an area is a function of it's employment base, resource mix, and relative advantages. To thrive locally, new employment should offer a better alternative to employment opportunities in the area, both current and of the recent past. In addition, for new employment to be attracted to these areas they need to offer a location package preferable to alternative sites. Analysis of the transferability of the "Lowell model" of reindustrialization to other economically depressed communities also highlights a key role and challenge for public policy: eliminating the years, possibly decades, of economic stagnation in an area between the loss of substantial numbers of jobs in traditional employment sectors and the entry of jobs in newer industries.

SUMMARY

Analyses of change both in the workplace and in the community suggest that public policies to facilitate technological change should focus on the shift in the provision of skills transferable among workplaces from the firm to the schools, worker mobility through retraining programs and relocation assistance, and employment diversity in local communities. In summary, public policies for integrating technological change in the workplace and in the community should incorporate an understanding of both the characteristics of the firm or area involved, and the dynamics of technological and production processes.

NOTES

Chapter 1: Introduction

1. See, for example, James R. Bright, *Automation and Management* (Boston: Harvard University Graduate School of Business Administration, 1958); Philip Kraft, *Programmers and Managers: The Routinization of Programming in the United States*, (New York: Springer-Verlag, 1977); Joan Greenbaum, *In the Name of Efficiency: A Study of Change in Data Processing Work* (Philadelphia: Temple University Press, 1979); Harry Braverman, *Labor and Monopoly Capital: The Degradation of Work in the Twentieth Century* (New York: Monthly Review, 1974); Heather Menzies, *Women and the Chip: Case Studies of the Effects of Informatics on Employment in Canada* (Montreal: Institute for Research on Public Policy, 1981); Jon Shepard, *Automation and Alienation: A Study of Office and Factory Workers* (Cambridge: The MIT Press, 1971).

2. See, for example, U.S. Congress, Office of Technology Assessment (OTA), *Technology and Structural Unemployment: Reemploying Displaced Adults* (Washington, D.C.: U.S. Government Printing Office, 1986); Paul Attewell and James Rule, "Computing and Organizations: What We Know and What We Don't Know," *Communications of the ACM* 27, no. 12, (December 1984): 1184–1192; Paul Adler, "Rethinking the Skill Requirements of New Technologies," Working Paper 9-784-076 (Boston, Mass.: Harvard University Graduate School of Business Administration, 1983); Stephen G. Peitchinis, *Computer Technology and Employment* (London: Macmillan, 1983.)

3. See, for example, Edwin E. Mansfield, *The Economics of Technological Change* (New York: W.W. Norton, 1968) pp. 134–145; Charles Killingsworth, "The Automation Story: Machines, Manpower and Jobs," in C. Markhan, *Jobs, Men and Machines* (New York: 1964), pp. 15–87; Peitchinis, *Computer Technology and Employment*, pp. 91–131; Attewell and Rule, "Computing

and Organizations"; Walter Buckingham, *Automation: Its Impact on Business and People* (New York: Harper and Brothers Publishers, 1961); Bright, *Automation and Management*; OTA, *Technology and Structural Unemployment.*

4. Mansfield, *Economics of Technological Change*, pp. 136–138; Peitchinis, *Computer Technology and Employment*, pp. 132–151; U.S. Congress, Office of Technology Assessment (OTA), *Computerized Manufacturing Automation, Employment, Education and the Workplace* (Washington, D.C.: U.S. Government Printing Office, April 1984), p. 103; Organization of Economic Cooperation and Development (OECD), *The Requirements of Automated Jobs* (Paris, 1965); Ewan Clague, "Effects of Technological Change on Occupational Employment," in OECD, *Requirements of Automated Jobs*, pp. 103–159; Morris Horowitz and Irwin Herrnstadt, "Changes in Skill Requirements of Occupations in Selected Industries," in *Report of the National Commission on Technology, Automation, and Economic Progress*, Vol. 1 (Washington, D.C.: U.S. Government Printing Office, 1966).

5. See, for example, Charles R. Walker, *Toward the Automatic Factory: A Case Study of Men and Machines* (New Haven, Conn.: Yale University Press, 1957); Charles R. Walker, "Changing Character of Human Work under the Impact of Technological Change," in *Report of the National Commission on Technology, Automation and Economic Progress*; Mansfield, *Economics of Technological Change*, pp. 134–143; Enid Mumford and Olive Banks, *The Computer and the Clerk* (London: Routledge and Kegan Paul, 1967); Arnold Weber, "Variety in Adaptation to Technological Change," in OECD, *Requirements of Automated Jobs*; Jack Steiber, "Manpower Adjustments to Automation and Technological Change in Western Europe," in *Report of the National Commission on Technology, Automation and Economic Progress*; Larry M. Blair, "Worker Adjustment to Changing Technology: Techniques, Processes, and Policy Considerations," in Eileen L. Collins and Lucretia Dewey Tanner, eds., *American Jobs and the Changing Industrial Base*, (Cambridge, Mass.: Ballinger, 1984), pp. 207–252.

6. Sar Levitan and Harold Sheppard, "Technological Change and the Community," in G. Somers, E. Cushman, and N. Weinberg, eds., *Adjusting to Technological Change* (New York: Harper and Row, 1963); G.F. Summers, S.D. Evans, F. Clemente, E.M. Beck, J. Minkoff, and E. Elwood, *Industrial Invasion of Non-Metropolitan America: A Quarter Century of Experience* (New York: Praeger, 1976).

7. Barry Bluestone and Bennett Harrison, *The Deindustrialization of America*, (New York: Basic Books, 1982); OTA, *Technology and Structural Unemployment; Report of the National Commission on Technology, Automation and Economic Progress*; A.T. Thwaites and R.P. Oakey, *The Regional Economic Impact of Technological Change* (New York: St. Martin's Press, 1985); Edward J. Malecki, "High Technology Sectors and Local Economic Development," in Edward M. Bergman, ed., *Local Economies in Transition* (Durham, N.C.: Duke University Press, 1986), pp. 129–142; Candee Harris, "Establishing High Technology Enterprises in Metropolitan Areas," in Bergman, *Local Economies in Transition*, pp. 165–184; Edward J. Malecki, "Technology and

Regional Development: A Survey," *International Regional Science Review* 8, no. 2 (1983): 89–125.

8. Segal Quince Wickstead, *The Cambridge Phenomenon: The Growth of High Technology Industry in a University Town* (Cambridge, England: Segal Quince Wickstead, 1985); Patricia M. Flynn, "Lowell: A High Technology Success Story," *New England Economic Review* (September/October 1984): 39–49; Summers, *Industrial Invasion of Non-Metropolitan America.*

9. U.S. Congress, Office of Technology Assessment (OTA), *Technology, Innovation and Regional Economic Development; Census of the State Government Initiatives for High Technology Industrial Development* (Washington, D.C.: U.S. Government Printing Office, 1983); Monroe W. Karmin, "High Tech: Blessing or Curse?" *U.S. News and World Report* 96, no. 2 (January 16, 1984): 38–44; "America Rushes to High Tech for Growth," *Business Week*, March 28, 1983, p. 85; D.L. Koch, W.N. Cox, D.W. Steinhauser, and P.V. Whigham, "High Technology: The Southeast Reaches out for Growth Industry," *Economic Review* 68, no. 9 (September 1983): 4–19.

10. Most definitions of "high technology" refer, at a minimum, to the following industries: drugs, office and computing machines, communications equipment, electronics components and accessories, engineering and scientific instruments, measuring and controlling devices, optical instruments and lenses, medical instruments and supplies, and photographic equipment and supplies. Frequently mentioned are the rest of the instruments industry, the chemical and electrical equipment industries, ordnance, miscellaneous transportation equipment, the aircraft industry, and various service-sector industries.

11. OTA, *Technology Innovation and Regional Economic Development*, pp. 5–25; Robert Vinson and Paul Harrington, *Defining High Technology Industries in Massachusetts* (Boston: Department of Manpower Development, 1979); Richard W. Riche, Daniel E. Hecker, and John U. Burgan, "High Technology Today and Tomorrow: A Small Slice of the Employment Pie," *Monthly Labor Review* 106, no. 11 (November 1983): 50–58; Helen Munzer and John Doody, *High Technology Employment: Massachusetts and Selected States* (Boston: Division of Employment Security, Job Market Research Research, March 1981); Lynn E. Browne, "Can High Tech Save the Great Lake States?" *New England Economic Review* (November/December 1983): 19–33; Karmin, "High Tech: Blessing or Curse?"; Donald Tomaskovic-Derey and S.M. Miller, "Can High-Tech Provide the Jobs?" *Challenge* 26, no. 2 (May/June 1983): 57–63; John M.L. Gruenstein, "Targeting High Technology in the Delaware Valley," *Business Review* (May/June 1984): 3–14; R. Greene, P. Harrington, and R. Vinson, "High Technology Industry: Identifying and Tracking an Emerging Source of Employment Strength," *New England Journal of Employment and Training* (Fall 1983); Edward J. Malecki, "High Technology and Local Economic Development," *Journal of the American Planning Association* 50, no. 3 (Summer 1984): 262–269.

12. Peter B. Doeringer and Patricia Flynn, "Manpower Strategies for Growth and Diversity in New England's High Technology Sector," in John C. Hoy

and Melvin H. Bernstein, eds., *New England's Vital Resource: The Labor Force* (Washington, D.C.: American Council on Education, 1982), pp. 11–35; Ann R. Markusen, "Defense Spending and the Geography of High-Tech Industries," in John Rees, ed., *Technology, Regions and Policy* (Totowa, N.J.: Rowman and Littlefield, 1986), pp. 94–119.

13. Doeringer and Flynn, "Manpower Strategies for Growth"; Robert Premus, *Location of High Technology Firms and Regional Economic Development* (Washington, D.C.: Joint Economic Committee of Congress, June 1982); Harris, "Establishing High Technology Enterprises"; Malecki, "High Technology Sectors"; Catherine Armington, Candee Harris, and Marjorie Odle, "Formation and Growth in High Technology Businesses: A Regional Assessment" Appendix B, in OTA, *Technology, Innovation and Regional Development*; Ann R. Markusen, "High Tech Jobs, Markets and Economic Development Prospects: Evidence from California," in Peter Hall and Ann R. Markusen, eds., *Silicon Landscapes* (Boston: Allen and Unwin, 1985) pp. 35–48; Massachusetts Division of Employment Security, "High Technology's Impact on the Massachusetts Economy Since 1976" (Boston: Division of Employment Security, November 1985); Robert Howard, "Second Class in Silicon Valley," *Working Papers* (September/October 1981): 21–31; Bennett Harrison, "Rationalization, Restructuring and Industrial Reorganization of Older Regions: The Economic Transformation of New England since World War II," Working Paper No. 72 (Cambridge, Mass.: Joint Center for Urban Studies of MIT and Harvard University, February 1982); Martin Carnoy, "High Technology and International Labour Markets," *International Labour Review* 124, no. 6 (November/December 1985): 643–660; M.A. Weiss, "High Technology Industries and the Future of Employment," *Built Environment* 9, no. 1 (1983): 51–60; P. Hall, A. Markusen, R. Osborn, and B. Wachsman, "The American Computer Software Industry: Economic Development Prospects," *Built Environment* 9, no. 1 (1983): 29–39.

14. See Elizabeth L. Useem, *Low Tech Education in a High Tech World* (New York: The Free Press, 1986); Henry Levin and Russell Rumberger, *The Education Implications of High Technology* (Palo Alto, Calif.: Institute for Research on Educational Finance and Governance, Stanford University, 1983); Office of Technology Assessment, *Automation and the Workplace: Selected Labor, Education and Training Issues* (Washington, D.C.: U.S. Government Printing Office, 1983); Educational Research Services, "Education for a High Technology Future: The Debate over the Best Curriculum" (Arlington, Va., May 1983); Stephen G. Peitchinis, *The Effect of Technological Changes on Educational and Skill Requirements of Industry* (Ottawa: Department of Industry, Trade and Commerce, 1978); Angelo C. Gilli, Sr., "Vocational Education and High-Technology," *Journal of Studies in Technical Careers* 6, no. 3 (Summer 1984): 187–197.

15. Harris, "Establishing High Technology Enterprises"; Amy K. Glasmeier, Peter Hall, and Ann R. Markusen, "Recent Evidence on High Technology Industries' Spatial Tendencies: A Preliminary Investigation," Working

Paper No. 417 (Berkeley: Institute of Urban and Regional Development, University of California-Berkeley, October 1983); Browne, "Can High Tech Save?"; Malecki, "High Technology Sectors"; Henry Levin and Russell Rumberger, "The Low-Skill Future of High Tech," *Technology Review* 86, no. 6 (August/September 1983): 18–21; Ray Oakey, *High Technology Small Firms* (New York: St. Martin's Press, 1984); OTA, *Technology, Innovation and Regional Economic Development*, p. 22; Armington, Harris, and Odle, "Formation and Growth."

16. Ann R. Markusen, *Profit Cycles, Oligopoly and Regional Development* (Cambridge, Mass.: The MIT Press, 1985); p. 44; Benjamin Stevens and Carolyn Brackett, *Industrial Location: A Review and Annotated Bibliography of Theoretical, Empirical and Case Studies* (Philadelphia: Regional Science Research Institute, 1967); OTA, *Technology, Innovation and Regional Economic Development*; Malecki, "High Technology Sectors," pp. 28, 138; Gruenstein, "Targeting High Technology."

17. Edward J. Malecki, "Corporate Organization of R & D and the Location of Technological Activities," *Regional Studies* 14 (1980): 219– 234; James Botkin, Dan Dimancescu, and Ray Stata, *Global Stakes: The Future of High Technology in America* (Cambridge, Mass.: Ballinger, 1982); Oakey, *High Technology Small Firms*, pp. 42–53; Thwaites and Oakey, *The Regional Economic Impact* pp. 6–7; Christopher Freeman, *The Economics of Industrial Innovation* (Cambridge, Mass.: The MIT Press, 1982); Malecki, "High Technology and Local Economic Development"; Carnoy, "High Technology and International Labour Markets."

18. John S. Hekman, "The Future of High Technology Industry in New England: A Case Study of Computers," *New England Economic Review* (January/February 1980): 5–17; Premus, *Location of High Technology Firms*; AnnaLee Saxenian, "The Urban Contradictions of Silicon Valley: Regional Growth and Restructuring of the Semiconductor Industry," *International Journal of Urban and Regional Research* 7 (1983): 237–261; Armington, Harris, and Odle, "Formation and Growth"; OTA, *Technology, Innovation and Regional Economic Development*, p. 3.

19. Russell Rumberger, "Changing Skill Requirements of Jobs in the U.S. Economy," *Industrial Labor Relations Review* 34, no. 4 (1981): 578–590; Abram J. Jaffe and Joseph Froomkin, *Technology and Jobs: Automation in Perspective* (New York: Praeger, 1966); Horowitz and Herrnstadt, "Changes in Skill Requirements"; P. Cain and D. Treiman, "The D.O.T. as a Source of Occupational Data," *American Sociological Review* 46, no. 3 (1981): 235–278; Attewell and Rule, "Computing Organizations."

20. The need to look at firm-level activities in order to understand regional economic development has been highlighted for over a decade in the literature. See, for instance, Morgan D. Thomas, "Growth Pole Theory, Technological Change and Regional Economic Development," *Papers of the Regional Science Association* 34 (1975): 3–25; R.R. Nelson and S.G. Winter, "Neoclassical vs. Evolutionary Theories of Economic Growth: Critique and Prospectus," *Economic Journal* 84 (1974): 886–905; Gunter Krumme

and Roger Hayter, "Implications of Corporate Strategies and Product Cycle Adjustments for Regional Changes," in Lyndhurst Collins and David F. Walker, eds., *Locational Dynamics of Manufacturing Activity* (New York: Wiley, 1975) pp. 325–356; Allan R. Pred, *City-Systems in Advanced Economies* (New York: Wiley, 1977). Also see Malecki,"Technology and Regional Development"; R.R. Nelson, "Research on Productivity Growth and Productivity Differences: Dead Ends and New Departures," *Journal of Economic Literature* 19, no. 3 (1981): 1029–1064.

21. Malecki, "Technology and Regional Development"; Krumme and Hayter, "Implications of Corporate Strategies", Thomas, "Growth Pole Theory."

22. In-depth studies of local labor markets have provided considerable insight into the functioning of labor markets and the impacts of workplace practices on the community. See, for example, F. Theodore Malm, "Recruiting and the Functioning of Labor Markets," *Industrial Labor Relations Research* 7, no. 4 (July 1954): 507–525 [San Francisco Bay Area]; Lloyd G. Reynolds, *The Structure of Labor Markets: Wages and Labor Mobility in Theory and Practice* (New York: Harper and Brothers, 1951) [New Haven]; Charles A. Myers and George P. Schultz, *The Dynamics of a Labor Market: A Study of the Impact of Employment Changes in Labor Mobility, Job Satisfactions and Company and Union Policies*, (New York: Prentice-Hall, 1951) [Nashua]; Krumme and Hayter, "Implications of Corporate Strategies" [Seattle]; Peter B. Doeringer, David G. Terkla, and Gregory Topakian, *Invisible Factors and Local Economic Development* (New York: Oxford University Press, forthcoming) [Fitchberg].

23. Segal Quince Wickstead, *The Cambridge Phenomenon*; Flynn, "Lowell: A High Technology Success Story."

Chapter 2: Production Life Cycles and Human Resources

1. Joel Dean, "Pricing Policies for New Products," *Harvard Business Review* 28, no. 6 (November 1950): 45–53. Formulation of an s-shaped growth for products and ideas has been credited to Gabriel Tarde, a French sociologist, in a work published in 1890; see J.J. van Duijn, *The Long Wave in Economic Life* (London: George Allen and Unwin, 1983), p. 21.

2. Theodore Levitt, "Exploit the Product Life Cycle," *Harvard Business Review* 43, no. 6 (November/December 1965): 81–94; Raymond Vernon, "International Investment and International Trade in the Product Cycle," *Quarterly Journal of Economics* 80, no. 2 (May 1966): 190–207; Louis T. Wells, Jr., ed., *The Product Life Cycle and International Trade* (Cambridge: Harvard University Press, 1972); Chester R. Wasson, *Dynamic Competitive Strategy and Product Life Cycles*, 3d ed. (Austin: Austin Press, 1978); Michael E. Porter, *Competitive Strategy* (New York: Free Press, 1980): Chap. 8.

3. Wasson, *Dynamic Competitive Strategy*; Nariman K. Dhalla and Sonia Yuspeh, "Forget the Product Life Cycle Concept," *Harvard Business Review*

54, no. 1 (January/February 1976): 102–112; van Duijn, *Long Wave*, pp. 22–26; William J. Abernathy, Kim B. Clark, and Alan M. Kantrow, *Industrial Renaissance* (New York: Basic Books, 1983); Porter, *Competitive Strategy*, pp. 238–245; W.B. Walker, *Industrial Innovation and International Trading Performance* (Greenwich, Conn.: JAI Press, 1979).

4. William J. Abernathy and James M. Utterback, "A General Model," in W.J. Abernathy *The Productivity Dilemma* (Baltimore: Johns Hopkins Press, 1978), ch. 4, pp. 68–84; William J. Abernathy and James M. Utterback, "Patterns of Industrial Innovation," in Robert R. Rothberg, ed., *Corporate Strategy and Product Innovation*, 2d ed. (New York: The Free Press, 1981), pp. 428–436; James M. Utterback and William J. Abernathy, "A Dynamic Model of Process and Product Innovation" 3, no. 6 *Omega* (1975): 639–656; Robert H. Hayes and Steven C. Wheelwright, "Link Manufacturing Process and Product Life Cycles," *Harvard Business Review* 57, no. 1 (January/February 1979): 133– 140; Robert H. Hayes and Steven C. Wheelwright, "The Dynamics of Process-Product Life Cycles," *Harvard Business Review* 57, no. 2 (March/April 1979): 127–136; Roy Rothwell and Walter Zegveld, *Reindustrialization and Technology* (New York: M.E. Sharpe, 1985).

5. Vernon, "International Investment"; Raymond Vernon, ed., *The Technology Factor in Intermediate Trade* (New York: Columbia University Press, 1970); Raymond Vernon, "The Product Cycle Hypothesis in a New International Environment," *Oxford Bulletin of Economics and Statistics* 41, no. 4 (1979): 255–267; Louis T. Wells, Jr., "International Trade: The Product Life Cycle Approach," in Louis T. Wells, Jr., ed., *Product Life Cycle*, pp. 3–33; Seev Hirsch, *Location of Industry and International Competitiveness* (Oxford, England: The Clarendon Press, 1967).

6. In general, the time period involved is shorter where economies of scale are significant at low production levels, where tariffs or transportation costs are high, when income elasticity of demand for the product is low, and where the size of the foreign market is large. See Wells, "International Trade," p. 13.

7. For examples of these patterns of dispersion see, Seev Hirsch, "The United States Electronics Industry in International Trade," in Louis T. Wells, Jr., ed., *The Product Life Cycle and International Trade* (Cambridge: Harvard University Press, 1972), pp. 39–52; John E. Tilton, *International Diffusion of Technology: The Case of Semiconductors* (Washington, D.C.: The Brookings Institution, 1971); Robert Stobough, "The Neotechnology Account of International Trade: The Case of Petrochemicals," in Louis T. Wells, Jr., ed., *The Product Life Cycle and International Trade* (Cambridge: Harvard University Press, 1972), pp. 83–105; John Hekman, "The Future of High Technology Industry in New England: A Case Study of Computer," *New England Economic Review* (January/February 1980): 5–17; John S. Hekman, "The Product Cycle and New England Textiles," *Quarterly Journal of Economics* 94, no. 4 (June 1980): 697–717; AnnaLee Saxenian, "Silicon Chips and Spatial Structure: The Industrial Basis of Urbanization in Santa Clara County, California," Working Paper 345 (Berkeley: Institute of

Urban and Regional Development, University of California, March 1981);
E.B. Alderfer and H.E. Michl, *Economics of American Industry* (New York:
McGraw-Hill, 1942).

8. Recent changes in the international environment, centered around the
increasing role of multinational, multiproduct firms in the world market,
are expected to affect the international distribution and dispersion patterns of employment. Expanding market sizes and opportunities for
worldwide standardization of products — and a lesser uncertainty about
demand and resource conditions — for example, will affect the size and
timing of various product cycles. In addition, the growth of conglomerates
suggests a lesser role for agglomeration economies in the location decisions of firms. Branch plants of large firms, for instance, may be subsidized
or supported through interplant transfers of personnel or other resources,
thus making it less vital for them to be located near other firms in similar
fields.

However, while the increasing role of multinational, multiproduct
firms suggests some modifications to the international product life-cycle
model as originally formulated, levels of risk, standardization of product
and equipment, and product demand — the key features underlying the
dynamics of the life-cycle models — continue to play critical roles in
determining production and employment patterns. See Raymond Vernon, "The Product Cycle Hypothesis in a New International Environment"; Gunter Krumme and Roger Hayter, "Implications of Corporate
Strategies and Product Cycle Adjustments for Regional Employment
Changes," in Lyndhurst Collins and David Walker, eds., *Location Dynamics of Manufacturing Activities*, (New York: Wiley, 1975), pp. 325–356.

9. John Rees and Howard Stafford, "High Technology Location and Regional Development: The Theoretical Base," in Office of Technology
Assessment (OTA), *Technology, Innovation and Regional Economic Development* (Washington, D.C.: U.S. Government Printing Office, July 1984),
Appendix A; Ann R. Markusen, *Profit Cycles, Oligopoly and Regional Devel-
opment* (Cambridge, Mass.: The MIT Press, 1985); Krumme and Hayter,
"Implications of Corporate Strategies"; R.D. Norton and John Rees, "The
Product Cycle and the Spatial Decentralization of American Manufactur-
ing," *Regional Studies* 13, no. 2 (1979): 141–151; Edward J. Malecki, "Tech-
nology and Regional Development: A Survey," *International Regional
Science Review* 8, no. 2 (1983): 89–125; Morgan D. Thomas, "Regional
Economic Development and the Role of Innovation and Technological
Change," in A.T. Thwaites and R.P. Oakey, eds., *The Regional Economic
Impact of Technological Change* (New York: St. Martin's Press, 1985), pp.
13–35; Morgan D. Thomas, "Growth Pole Theory, Technological Change,
and Regional Economic Development," *Papers of the Regional Science
Association* 34 (1975): 3–25.

10. Edward J. Malecki, "High Technology and Local Economic Development," *Journal of the American Planning Association* 50, no. 3 (Summer 1984):
262–269; Edward J. Malecki, "Firm Size, Location and Industrial R & D: A

Disaggregated Analysis," *Review of Business and Economic Research* 16, no. 2 (1980): 29–42; Edward J. Malecki, "Corporate Organization of R & D and the Location of Technological Activities," *Regional Studies* 14 (1980): 219–234; Rees and Stafford, "High Technology Location," pp. 101–102; Hekman, "Future of High Technology Industry."

11. G.L. Clark, "The Employment Relation and Spatial Division of Labor: A Hypothesis," *Annals of the Association of American Geographers* 71 (1981): 412–424; Doreen Massey, "In What Sense a Regional Problem?" *Regional Studies* 13, no. 2 (1979): 231–241; R.A. Erickson and T.R. Leinbach, "Characteristics of Branch Plants Attracted to Nonmetropolitan Areas," in R.E. Lonsdale and H.L. Seyler, eds., *Nonmetropolitan Industrialization* (New York: Winston/Wiley, 1979) pp. 57–78; Hekman, "Future of High Technology Industry"; Hekman, "Product Cycle"; Markusen, *Profit Cycles*; Martin Carnoy, "High Technology and International Labour Markets," *International Labour Review* 124, no. 6 (November/December 1985): 643–660; Robyn Swaim Philips and Avis C. Vidal, "Restructuring and Growth Transitions of Metropolitan Economies," in Edward M. Bergman, ed., *Local Economies in Transition* (Durham, N.C.: Duke University Press, 1986), pp. 59–83; John S. Hekman, "Can New England Hold onto Its High Technology Industry?" *New England Economic Review* (March/April 1980): 35–44; P. Hall, A. Markusen, R. Osborne, and B. Wachsman, "The American Computer Software Industry: Economic Development Prospects," *Built Environment* 9, no. 1 (1983): 29–39; Stephen G. Peitchinis, *Computer Technology and Employment* (London: Macmillan, 1983), pp. 21–22.

12. Tilton, *International Diffusion*; Hirsch, *Location of Industry*; U.S. Congress, Office of Technology Assessment (OTA), *International Competitiveness in Electronics* (Washington, D.C.: U.S. Government Printing Office, November 1983).

13. Saxenian, "Silicon Chips"; Tilton, *International Diffusion*; James W. Harrington, Jr., "Learning and Locational Change in the American Semiconductor Industry," in John Rees, ed., *Technology, Regions and Policy* (Totowa, N.J.: Rowman and Littlefield, 1986), pp. 120–137; Kenneth Flamm, "Internationalization in the Semiconductor Industry," in Joseph Grunwald and Kenneth Flamm, eds., *The Global Factory* (Washington, D.C.: The Brookings Institutions, 1985), pp. 38–136.

14. Hekman, "Future of High Technology Industry."

15. Hekman, Ibid.; Robert Premus, *Location of High Technology Firms and Regional Economic Development*, (Washington, D.C.: Joint Economic Committee of Congress, June 1982); Sarah Kuhn, *Computer Manufacturing in New England* (Cambridge, Mass.: Joint Center for Urban Studies of MIT and Harvard University, April 1982).

16. Hekman, "Can New England Hold?"

17. David Ford and Chris Ryan, "Taking Technology to Market," *Harvard Business Review* 59, no. 2 (March/April 1981): 117–126; Richard N. Foster, "A Call for Vision in Managing Technology," *Business Week*, May 24, 1982, pp. 24, 26, 28, 33; William L. Shanklin and John K. Ryans, Jr., *Marketing*

High Technology (Lexington, Mass.: Lexington Books, 1984); Walter Kiechell III, "The Decline of the Experience Curve," *Fortune*, October 5, 1981, pp. 139–140, 144, 146; Arnold C. Cooper and Daniel Schendel, "Strategic Responses to Technological Threats, *Business Horizons* 19, no. 1 (February 1976): 61–69; Richard N. Foster, "To Exploit New Technology, Know When to Junk the Old," *Wall Street Journal* 2 May 1983, p. 22.

18. Shanklin and Ryans, *Marketing High Technology*, pp. 106–107; Ford and Ryan, "Taking Technology to Market," pp. 117–118.

19. Edwin Mansfield, *Industrial Research and Technological Innovation* (New York: W.W. Norton and Company, 1968); Mansfield, *Economics of Technological Change*, pp. 119–133; Christopher Freeman, *The Economics of Industrial Innovation*, 2d ed. (Cambridge, Mass.: The MIT Press, 1982); National Science Foundation (NSF), *The Process of Technological Innovation: Reviewing the Literature* (Washington, D.C.: NSF, May 1983); Nathan Rosenberg, *Inside the Black Box: Technology and Economics* (Cambridge, England: Cambridge University Press, 1982) pp. 55–162; Joseph Schumpeter, *Capitalism, Socialism and Democracy* (New York: Harper and Row, 1942).

20. Nathan Rosenberg, *Perspectives on Technology*, (Armonk, N.Y.: M.E. Sharpe 1976), pp. 141–211; Stephen G. Peitchinis, *Computer Technology and Employment* (London: Macmillan, 1983), pp. 21–26; E. Mansfield, J. Rapoport, A. Romeo, E. Villani, S. Wagner, and F. Husic, *The Production and Application of New Industrial Technology* (New York: W.W. Norton and Company, 1977), pp. 68–86, 126–143; Mansfield, *The Economics of Technological Change*, pp. 99–125.

21. Edgar M. Hoover, *The Location of Economic Activity* (New York: McGraw-Hill, 1948), pp. 174–175; Markusen, *Profit Cycles*; Hekman, "Future of High Technology Industry"; Freeman, *Economics of Industrial Innovation*; Malecki, "High Technology and Local Development"; Peitchinis, *Computer Technology*, pp. 103–106; Ray Oakey, *High Technology Small Firms* (New York: St. Martin's Press, 1984), pp. 39– 40; Rees and Stafford, "High Technology Location," pp. 97–100; J.E. Browning, *How to Select a Business Site* (New York: McGraw-Hill, 1980).

22. James R. Bright, *Automation and Management* (Boston: Harvard University Graduate School of Business Administration, 1958); Philip Kraft, *Programmers and Managers: The Routinization of Programming in the United States* (New York: Springer-Verlag, 1977); Joan Greenbaum, *In the Name of Efficiency: A Study of Change in Data Processing Work* (Philadelphia: Temple University Press, 1979); Harry Braverman, *Labor and Monopoly Capital: The Degradation of Work in the Twentieth Century* (New York: Monthly Review, 1974); Heather Menzies, *Women and the Chip: Case Studies of the Effects of Informatics on Employment in Canada* (Montreal: Institute for Research on Public Policy, 1981); William J. Abernathy and James M. Utterback, "Patterns of Industrial Innovation," in Robert R. Rothberg, ed., *Corporate Strategy and Product Innovation*, 2d ed. (New York: The Free Press, 1981), pp. 428– 436; James M. Utterback and William J. Abernathy, "A Dynamic Model of Process and Product Innovation," 3, no. 6 *Omega* (1975):

639–656; Kenneth J. Arrow, "The Economic Implications of Learning by Doing," *The Review of Economic Studies* 29, no. 1 (July 1962): 155–173; Hirsch, *Location of Industry*, chap. 2; Kuhn, *Computer Manufacturing*, pp. 103–105.

23. For a detailed description of the "deskilling process," an effect attributed to technological change by Adam Smith in the eighteenth century, see James R. Bright, "Does Automation Raise Skill Requirements?" *Harvard Business Review* 36, no. 4 (July–August 1958): 85–98.

24. Ibid.; Kraft, *Programmers and Managers*; Greenbaum, *In the Name of Efficiency*; Kuhn, *Computer Manufacturing*; Peitchinis, *Computer Technology*, pp. 100–101.

25. This section draws heavily upon Patricia M. Flynn, "Technological Change, the 'Training Cycle' and Economic Development," in John Rees, ed., *Technology, Regions and Policy* (Totowa, N.J.: Rowman and Littlefield, 1986), pp. 282–308. The 'training cycle' was originally introduced in Patricia M. Flynn, "Production Life Cycles and Their Implications for Education and Training," Final Report, Grant No. NIE-G- 82-0033 (Washington, D.C.: National Institute of Education, February 1984).

26. Michael J. Piore, "On-the-Job Training and Adjustment to Technological Change," *The Journal of Human Resources* 3, no. 4 (Fall 1968): 435–449; Patricia M. Flynn, "The Impact of Technological Change on Jobs and Workers," Final Report, Grant No. 21-25-82-16 (Washington, D.C.: U.S. Department of Labor, Employment and Training Administration, March 1985; Harold Goldstein and Bryna Shore Fraser, "Training for Work in the Computer Age: How Workers Who Use Computers Get Their Training," Research Reprint Series No. RR-85-09, National Commission on Employment Policy, June 1985.

27. For a detailed analysis of specific versus general skills, see Gary Becker, *Human Capital* (New York: Columbia University Press, 1964.)

28. Flynn, "Production Life Cycles"; Kraft, *Programmers and Managers*, pp. 104–106; Patricia M. Flynn, "Occupational Education and Training: Goals and Performance," in Bruce Vermeulen and Peter B. Doeringer, eds., *Jobs and Training: Vocational Policy and the Labor Market* (Boston: Martinus Nijhoff Publishing, 1981), pp. 50–71; Stephen J. Franchak, "Factors Influencing Vocational Education Program Decisions," in Robert E. Taylor, Howard Rosen, and Frank C. Pratzner, eds., *Responsiveness of Training Institutions to Changing Labor Market Demands* (Columbus, Oh.: The National Center for Research in Vocational Education, Ohio State University, 1983), pp. 267–293.

29. Peter Doeringer and Michael J. Piore, *Internal Labor Markets and Manpower Analysis* (Armonk, N.Y.: M.E. Sharpe, 1985); Paul Osterman, ed., *Internal Labor Markets* (Cambridge, Mass.: The MIT Press, 1984).

30. Bruce Vermeulen and Susan Hudson-Wilson, "The Impact of Workplace Practices on Education and Training Policy," in Vermeulen and Doeringer, eds., *Jobs and Training*; Peter B. Doeringer, *Workplace Perspectives on Education and Training* (Boston: Martinus Nijhoff Publishing, 1981;

Seymour Lusterman, *Education in Industry* (New York: The Conference Board, 1977); Doeringer and Piore, *Internal Labor Markets*; E.S. Stanton, *Successful Personnel Recruiting and Selection* (New York: AMACOM, 1977); J. Gaston, "Labor Market Conditions and Employer Hiring Standards," *Industrial Relations* 11, no. 2, (May 1972): 272–278; Anthony Carnevale and Harold Goldstein, *Employee Training: Its Changing Role and an Analysis of New Data* (Washington, D.C.: American Society for Training and Development, 1983); Richard Lester, *Hiring Practices and Labor Competition* (Princeton: Industrial Relations Section, Princeton University, 1954).

31. Ian Benson and John Lloyd, *New Technology and Industrial Change* (London: Kogan Page, 1983); Everett Rogers and Judith Laisen, *Silicon Valley Fever* (New York: Basic Books, 1984); OTA, *Technology, Innovation and Regional Economic Development*; Carnoy, "High Technology"; Peter S. Albin, "Job Design within Changing Patterns of Technological Development," in Eileen Collins and Lucretia Dewey Tanner, eds., *American Jobs and the Changing Industrial Base* (Cambridge, Mass.: Ballinger, 1984), pp. 125–162.

32. Peter B. Doeringer and Patricia Flynn, "Manpower Strategies for Growth and Diversity in New England's High Technology Sector," in John C. Hoy and Melvin H. Bernstein, eds., *New England's Vital Resource: The Labor Force* (Washington, D.C.: American Council on Education, 1982), pp. 11–35; Patricia M. Flynn, "Employer Response to Skill Shortages: Implications for Small Business," *Proceedings of the Small Business Research Conference* (Waltham, Mass.: Bentley College, 1981).

   More generally, small firms are said to be at a disadvantage regarding technological adoptions due to their limited financial resources and scale of production. See, for instance, Roy Rothwell and Walter Zegveld, *Innovation and the Small and the Medium Sized Firm* (Boston: Kluwer Nijhoff Publishing, 1981); Freeman, *Economics of Industrial Innovation*; Morton I. Kamien and Nancy L. Schwartz, *Market Structure and Innovation* (Cambridge, England: Cambridge University Press, 1982).

Chapter 3: Technological Change, Skills, and Jobs

1. The case studies in the data base were often conceived independently of one another, conducted by many different individuals and organizations and produced for a variety of reasons. They vary considerably in scope and depth. As such, inferences about the relative frequencies with which events occur cannot be drawn. The sample may not be used, for instance, to generate adoption rates of technologies by size of firm or by industry. Similarly, inferences cannot be drawn from such studies about the *extent* of layoffs, upgrading, deskilling, transfers, and other results of technological adoptions. Furthermore, one cannot use the sample to forecast the

number of jobs likely to be created or lost as a result of adoptions of future technologies.

2. Pennsylvania State Employment Service, *The Effect of Automation on Occupations and Workers in Pennsylvania* (May 1965), p. 49.

3. James R. Bright, *Automation and Management* (Boston: Harvard University Graduate School of Business Administration, 1958) p. 175.

4. E.P. Mathias, "Management of Technological Change in Railways," in C.P. Thakur and G.S. Aurora, *Technological Change and Industry* (New Delhi: Shri Ram Centre for Industrial Relations, 1971) pp. 111, 115.

5. Bright, *Automation and Management*, p. 160.

6. "New Steelmaking Techniques and Computerized Control," in Solomon Barkin, ed., *Technological Change and Manpower Planning* (Paris: Organization of Economic Cooperation and Development, 1967) p. 219.

7. Bright, *Automation and Management*, p. 193.

8. B.R. Paul, "The Introduction of Electronic Data Processing in Life Assurance," *Personnel Practice Bulletin* 18, no. 2 (June 1962): 9.

9. Edward B. Jakubauskas, "Adjustment to an Automatic Airline Reservation System," Report 137 (Washington, D.C.: U.S. Department of Labor, Bureau of Labor Statistics, 1958), p. 95.

10. Barry Wilkinson, "Managing with New Technology," *Management Today* (October 1982): 33.

11. Ibid., p. 34.

12. Ibid., p. 34.

13. See, for example, A. Willener, "A French Steelworks — Closure of an Old Rolling Mill: Problems of Co-ordinated Transfers of Personnel," in Organization of Economic Cooperation and Development (OECD), *Adjustment of Workers to Technological Change at the Plant Level*, Supplement to the Final Report (Paris, 1966), pp. 145–158.

14. Stanford Research Institute, *Management Decisions to Automate*, adapted by Harry F. Bonfils (Washington, D.C.: U.S. Department of Labor, Office of Manpower, Automation and Training, 1965), pp. 21–22.

15. Sir Stuart Mitchell, "Planning Coordination: British Railways," in OECD, *Adjustment of Workers to Technological Change at the Plant Level*, Supplement to the Final Report (Paris, 1966), pp. 184–196.

16. A. Willener, "Manpower Planning and Redeployment in an Expanding Business," in OECD, *Adjustment of Workers to Technological Change at the Plant Level*, Supplement to the Final Report (Paris, 1966), pp. 170–171.

17. Mathias, "Management of Technological Change in Railways," pp. 101–133.

18. "Modernization and Staff Reduction in a Dyestuffs Plant," in Solomon Barkin, ed., *Technological Change and Manpower Planning* (Paris: Organization of Economic Cooperation and Development, 1967), pp. 242–248.

19. Edward R.F.W. Crossman and Stephen Laner, *The Impact of Technical Change on Manpower Skill Demand: Case-Study Data and Policy Implications* (Berkeley: University of California, 1969), p. 108.

20. "ECM Engineering Designs," in Incomes Data Services, Ltd., "Introducing New Technology," IDS Study 202 (London, June 1980), p. 11.
21. "British Sugar Corporation," in Incomes Data Services, Ltd., "Introducing New Technology," IDS Study 202 (London, June 1980), pp. 8–9.
22. Bright, *Automation and Management*, pp. 177–178.
23. Ibid., p. 178.
24. Arkadii Erivansky, *A Soviet Automatic Plant* (Moscow: Foreign Language Publishing House, 1955).
25. Stanford Research Institute, *Management Decisions to Automate*, pp. 25–26.
26. Pennsylvania State Employment Service, *The Effect of Automation*, p. 57.
27. U.S. Department of Labor, Bureau of Labor Statistics, "A Case Study of a Company Manufacturing Electronic Equipment," Studies of Technology, No. 1 (Washington, D.C., October 1955); K.G. van Auken, Jr., "Plant Level Adjustments to Technological Change," *Monthly Labor Review* 76, no. 4 (April 1953): 388–391.
28. Floyd C. Mann and Laurence K. Williams, "Observations on the Dynamics of a Change to Electronic Data Processing Equipment," *Administrative Science Quarterly* 5, no. 2 (September 1960): 253.
29. Russell Wilkins, *Microelectronics and Employment in Public Administration: Three Ontario Municipalities, 1976–1980* (Ontario: Ministry of Labor, July 1981) p. 23.
30. Heather Menzies, *Women and the Chip: Case Studies of the Effects of Informatics on Employment in Canada* (Montreal: Institute for Research on Public Policy, 1981) p. 37.
31. van Auken, "Plant Level Adjustments"; Stanford Research Institute, *Management Decisions to Automate*, pp. 22–24.
32. Wilkins, *Microelectronics and Employment*, p. 24.
33. See Menzies, *Women and the Chip*, pp. 30–31.
34. Ibid., pp. 34–35.
35. Ibid., p. 37.
36. See, for example, Menzies, Ibid., pp. 33-40. Joan Greenbaum provides an illustrative example of the deskilling process as it affects computer programming:

> Initially, programs had to be written in a detailed and complex format called machine language. This required a great deal of skill and knowledge of the machine, and was replaced by assembly language coding which simplified the instruction process by allowing the programmer to code fewer and less complicated instructions. By 1965 general purpose computers like IBM's could use more generalized instruction sequences; and easier-to-use languages like COBOL, a language for business processing, came into widespread use. These languages removed the programmer from the technical detail of the equipment and required only the ability to transcribe a given solution into an English-like series of instructions. The last development has been the introduction of pre-planned application languages, where programmers need only insert a prearranged series of codes. The development of pre-planned applications has resulted in the total removal of technical skills from some of the tasks of programming [*In the*

Name of Efficiency: A Study of Change in Data Processing Work (Philadelphia: Temple University Press, 1979), p. 48].

37. See Menzies, *Women and the Chip*; Russell Wilkins, *Microelectronics and Employment*.
38. Bright, *Automation and Management*, p. 185.
39. Mann and Williams, "Observations on the Dynamics," p. 253.
40. Harold Craig, "Administering Technological Change in a Large Insurance Office — A Case Study," *Industrial Relations Research Association Proceedings* (1954): 131, 133.
41. Wilkins, *Microelectronics and Employment*, p. 31.
42. Mann and Williams, "Observations on the Dynamics," p. 25.
43. "Case Study of Mechanization of Materials Handling," in U.S. Department of Labor, Bureau of Labor Statistics, *Impact of Technological Change and Automation in the Pulp and Paper Industry*, Bulletin No. 1347 (Washington, D.C.: U.S. Government Printing Office, October 1962), p. 43.
44. "The Introduction of Four-Shift Working in Paper Manufacture," in Barkin, *Technological Change and Manpower Planning*, p. 87.
45. Canadian Department of Labor, *Technological Changes and Skilled Manpower Electronic Data Processing Occupations in a Large Insurance Company* (Ottawa: Research Program in the Training of Skilled Manpower, 1961), p. 30.
46. See C.W. Phalen, "Automation and the Bell System," in Howard B. Jacobson and Joseph S. Roucek, eds., *Automation and Society* (New York: Philosophical Library, 1959); Roy Rothwell and Walter Zegveld, "National Coal Board," in Roy Rothwell and Walter Zegveld, eds., *Technical Change and Employment* (New York: St. Martin's Press, 1979), pp. 66– 76. Also see Walter Buckingham, *Automation: Its Impact on Business and People* (New York: Harper and Brothers Publishers, 1961), pp. 117– 119 for a discussion of "silent firing."
47. See, for example, Wilkins, *Microelectronics and Employment*.

Chapter 4: Staffing and Training Practices over the Technology Life Cycle

1. For earlier studies on employer adjustments to technological change, see Edwin Mansfield, *The Economics of Technological Change* (New York: W.W. Norton and Company, 1968) pp. 134–161; Michael J. Piore, "On- the-Job Training and Adjustment to Technological Change," *Journal of Human Resources* 3, no. 4 (Fall 1968), pp. 435–449; James R. Bright, *Automation and Management* (Boston: Harvard University Graduate School of Business Administration, 1958) pp. 123–131; Peter B. Doeringer and Michael J. Piore, *Internal Labor Markets and Manpower Analysis* (Armonk, N.Y.: M.E. Sharpe, 1985); K. van Auken, "Personnel Adjustment to Technological Change," in Howard B. Jacobson and Joseph S. Roucek, eds., *Automation and Society* (New York: Philosophical Library, 1959), pp. 387–391; Arnold Weber, "The

Interplant Transfer of Displaced Employees," in G. Somers, E. Cushman, and N. Weinberg, eds., *Adjusting to Technological Change* (New York: Harper and Row, 1963), pp. 95–143.

2. "The Effects of External Forces on Staff Planning," in Solomon Barkin, ed., *Technological Change and Manpower Planning* (Paris: Organization of Economic Cooperation and Development, 1967), p. 115.

3. Mariane B. Edson, "Managing the People and the System," *Word Processing, Selection, Implementation and Uses* (Uxbridge, England: Online Conference Limited, 1979), p. 114.

4. Russell Wilkins, *Microelectronics and Employment in Public Administration: Three Ontario Municipalities, 1976–1980* (Ontario: Ministry of Labor, July 1981), p. 40.

5. Frank H. Cassell, "Corporate Manpower Planning and Technical Change at the Plant Level, in Organization of Economic Cooperation and Development *Adjustment of Workers to Technical Change at the Plant Level, Supplement to the Final Report* (Paris, 1966), p. 126.

6. Russell Wilkins, *Microelectronics and Employment* , p. 10.

7. Ibid., p. 9.

8. G.P. Schultz and T.L. Whisler, eds., *Management, Organization and the Computer* (Glencoe, Ill.: The Free Press, 1960), pp. 203–204.

9. For a discussion of the decision to "make or buy" skilled workers in a rapidly changing labor market, see Thomas A. Barocci and Paul Cournoyer, "Make or Buy: Computer Professionals in a Demand-Driven Environment," Working Paper 1342 (Cambridge, Mass.: Sloan School of Management, MIT, September 1982); Thomas A. Barocci and Kirsten R. Wever, "Information Systems Careers and Human Resource Management," Working Paper 1482 (Cambridge, Mass.: Sloan School of Management, MIT, September 1983). Also see Paul Cournoyer, "Mobility of Information Systems Personnel: An Analysis of a Large Computer Firm's Experience" (Ph.D. dissertation, MIT, 1983); Rosabeth Moss Kanter, "Variations in Managerial Career Structures in High-Technology Firms: The Impact of Organizational Characteristics on Internal Labor Market Patterns," in Paul Osterman, ed., *Internal Labor Markets* (Cambridge, Mass.: The MIT Press, 1984), pp. 109–132.

10. Floyd C. Mann and Laurence K. Williams, "Observations on the Dynamics of a Change to Electronic Data Processing Equipment," *Administrative Science Quarterly* 5, no. 2 (September 1960): 241.

11. "Retraining of Aircraft Production Workers, Technicians and Engineers," in U.S. Department of Labor, Bureau of Labor Statistics, *Industrial Retraining Programs for Technological Change — A Study of the Performance of Older Workers*, Bulletin 1368 (Washington, D.C., June 1963), p. 13.

12. Cassell, "Corporate Manpower Planning," p. 126.

13. N.H. Greenhalgh, "Automation at Gawith's Bakery," *Personnel Practice Bulletin* 18, no. (June 1962): 29–30.

14. David G. Osborn, "Automatic Data Processing in the Large Company," in Howard Boone Jacobson and Joseph S. Roucek, eds., *Automation and*

Society (New York: Philosophical Library, 1959), p. 164.

15. Canadian Department of Labor, *Technological Change and Skilled Manpower Electronic Data Processing Occupations in a Large Insurance Company* (Ottawa: Research Program in the Training of Skilled Manpower, 1961), p. 28.

16. Ibid., p. 32.

17. Ibid., p. 3.

18. Osborn, "Automatic Data Processing," p. 10.

19. Susan Eckert, "Continuous Technological Advances Require Continuing Employee Education," *Telephony* (November 9, 1981): 26.

20. Wisconsin State Employment Service, "A Large Insurance Company Automates: Workforce Implications of Computer Conversion," Automation Manpower Services Program Demonstration Project No. 3, Madison, April 1964), p. 46.

21. Pamela Haddy, "Some Thoughts on Automation in a British Office," *Journal of Industrial Economics* 6 (1958): 161–170.

22. Sandra Salmans, "Pilkington's Progressive Shift," *Management Today* 163 (September 1980): 66–73.

23. Solomon Barkin, ed., *Technological Change and Manpower Planning* (Paris: Organization of Economic Cooperation and Development, 1967), pp. 36–45, 249–255; Stanford Research Institute, *Management Decisions to Automate*, adapted by Harry F. Bonfils, (Washington, D.C.: U.S. Department of Labor, Office of Manpower, Automation and Training, 1965), pp. 16–18; E.P. Mathias, "Management of Technological Change in Railways," in C.P. Thakur and G.S. Aurora, *Technological Change and Industry* New Delhi: Shri Ram Centre for Industrial Relations, 1971), pp. 101–133.

24. Ethel Best, "The Change from Manual to Dial Operation in the Telephone Industry," Bulletin No. 10, (Washington, D.C.: U.S. Women's Bureau, 1933); "Kaiser Steel Corporation and the U.S. Steelworkers of America," in Diebold Institute of Public Service Studies, *Labor Management Contracts and Technological Change* (New York: Praeger, 1969).

25. Wisconsin State Employment Services, "A Large Insurance Company Automates," p. 40.

26. Maryland Department of Employment Security, *The Impact of Technological Change in the Banking Industry* (Baltimore: Automation Manpower Services Program, June 1967).

27. Wilkins, *Microelectronics and Employment*, p. 43.

28. Salmans, "Pilkington's Progressive Shift."

29. Ibid., p. 69.

30. Ibid., p. 70.

31. These results are consistent with the literature on worker displacement more generally. See, U.S. Congress, Office of Technology Assessment, (OTA) *Technology and Structural Unemployment: Reemploying Displaced Adults* (Washington, D.C.: U.S. Government Printing Office, 1986). George Schultz and Arnold Weber, *Strategies for the Displaced Worker* (New York: Harper and Row, 1966); Kevin Hollenbeck, Frank C. Pratzner, and Howard Rosen, *Displaced Workers: Implications for Educational and Training*

Institutions (Columbus, Oh.: The National Center for Research in Vocational Education, 1984); Barry Bluestone, "Industrial Dislocation and Its Implications for Public Policy," in Kevin Hollenbeck, Frank C. Pratzner, and Howard Rosen, eds., *Displaced Workers: Implications for Educational and Training Institutions* (Columbus, Oh.: The National Center for Research in Vocational Education, 1984), pp. 45–68.

32. With the exception of the case in Harold L. Sheppard and James L. Stern, "Impact of Automation on Workers in Supplier Plants," *Labor Law Journal* 8, no. 10 (1957): 714–718, the case studies in this sample rarely addressed the issue of race.

33. Edson, "Managing the People," p. 108.

34. "Kaiser Steel Corporation and the U.S. Steelworkers of America," in Diebold Institute for Public Service Studies, *Labor Management Contracts and Technological Change* (New York: Praeger, 1969), p. 30.

35. Ibid., p. 33.

36. Otis Lipstrev and Kenneth A. Reed, *Transition to Automation: A Study of People, Production and Change* (Boulder: University of Colorado Press, 1964).

37. "No Shuttlecocks at Parlin," *Fortune*, February 1961, pp. 189–190.

38. Bright, *Automation and Management*, p. 178.

39. Ibid., p. 178.

40. For an historical look at the development of sex-typed jobs see Julie A. Matthaei, *An Economic History of Women in America* (New York: Schocken Books, 1982); Elizabeth Baker, *Technology and Women's Work* (New York: Columbia University Press, 1964); Ann H. Stromberg and Shirley Harkness, eds., *Women Working: Theories and Facts in Perspective* (Palo Alto, Calif.: Mayfield, 1978).

41. B.R. Paul, "The Introduction of Electronic Data Processing in Life Assurance," *Personnel Practice Bulletin* 18, no. 2 (June 1962): 7–11. Wisconsin State Employment Services, "A Large Insurance Company Automates"; Heather Menzies, *Women and the Chip: Case Studies of the Effects of Informatics on Employment in Canada* (Montreal: Institute for Research on Public Policy, 1981), pp. 33–40. Also see Barbara Baran, "Office Automation and Women's Work: The Technological Transformation of the Insurance Industry," in M. Castells, ed., *High Technology, Space and Society* (Beverly Hills: Sage Publications, 1985).

42. Paul, "The Introduction of Electronic Data Processing," p. 9.

43. Richard W. Riche and James R. Alliston, "Impact of Office Automation in the IRS," *Monthly Labor Review* 86, no. 4 (April 1963): 390.

44. For references of the problems encountered by older workers, more generally, see A. Ross and J. Ross, "Employment Problems of Older Workers," *Studies in Unemployment* (Washington, D.C.: Senate Special Committee on Unemployment Problems, 1959; Philip L. Rones, "The Labor Market Problems of Older Workers," *Monthly Labor Review* 106, no. 5 (May 1983): 3–12.

45. Greenhalgh, "Automation of Gawith's Bakery," p. 28.

46. Schultz and Whisler, *Management, Organization and the Computer*, p. 204.

47. Keith Dickson, "Petfoods by Computer: A Case Study of Automation," in Tom Forester, ed., *The Microelectronic Revolution* (Cambridge, Mass.: The MIT Press, 1980), p. 178.

48. Barkin, *Technological Change*, pp. 36–45; Ibid., pp. 249–255; Stanford Research Institute, *Management Decisions*, pp. 16–18; Mathias, "Management of Technical Change."

49. Menzies, *Women and the Chip*, p. 63.

50. Wilkins, *Microelectronics and Employment*, p. 27.

51. Menzies, *Women and the Chip*, p. 39.

Chapter 5: Technological Change and the Local Labor Market

1. "High Technology Makes Lowell a Model of Reindustrialization," *New York Times*, 10 August 1982, p. 1; "Lowell: From Riches to Rags and Back Again," *Dun's Review*, July 1980, pp. 38–39. Also see U.S. Congress, Office of Technology Assessment (OTA), *Technology and Structural Unemployment: Reemploying Displaced Adults*, (Washington, D.C.: U.S. Government Printing Office, 1986), p. 309; William Gorham, "Trade Winds and Urban Industrial Change" (Speech delivered at the International Symposium on Revitalization of Local Communities and the Role of Cities, Amagasaki, Japan, October 6, 1986 [available from The Urban Institute, Washington D.C.]).

2. This section draws heavily upon Patricia M. Flynn, "Technological Change, the 'Training Cycle,' and Economic Development," in John Rees, ed., *Technology, Regions and Policy* (Totowa, N.J.: Rowman & Cattlefield, 1986), pp. 286–293; and Patricia M. Flynn, "Lowell: A High Technology Success Story," *New England Economic Review* (September/October 1984): 39–49. Patricia M. Flynn, "Production Life Cycles and Their Implications for Education and Training," Final Report, Grant No. NIE G-82-0033-16, (Washington, D.C.: National Institute of Education, February 1984).

3. For more detail on Lowell's early history see Thomas Dublin, *Women at Work: The Transformation of Work and Communities in Lowell, Massachusetts 1826–1860* (New York: Columbia University Press, 1979); Arthur L. Eno, Jr., *Cotton Was King: A History of Lowell, Massachusetts* (Somersworth, N.H.: New Hampshire Publishing Company, 1976); George F. Kennegott, *The Record of a City: A Social Survey of Lowell, Massachusetts* (New York: Macmillan, 1912); H.C. Meserve, *Lowell: An Industrial Dream Come True* (Boston: National Association of Cotton Manufacturers, 1923) Margaret Terrell Parker, *Lowell: A Study of Industrial Development* (New York: Kennikat, 1940).

4. Eno, *Cotton Was King*, p. 256.

5. Massachusetts Division of Employment Security, ES-202 Survey data. These data account for approximately 97 percent of all wage and salary

employment in the state.

6. Based on the Massachusetts Division of Employment Security's definition of high-technology manufacturing employment, which includes the following twenty Standard Industrial Classification (SIC) three-digit-level industries: drugs (SIC 283), ordnance and accessories (SIC 348), office and computing machinery (SIC 357), electrical distribution equipment (SIC 361), electrical industrial apparatus (SIC 362), household appliances (SIC 363), electrical lighting and wiring (SIC 364), radio and television receiving equipment (SIC 365), communication equipment (SIC 366), electronic components and accessories (SIC 367), miscellaneous electrical equipment and supplies (SIC 369), guided missiles and space vehicles (SIC 376), miscellaneous transportation equipment (SIC 379), engineering and scientific instruments (SIC 381), measuring and controlling instruments (SIC 382), optical instruments and lenses (SIC 383), medical instruments and supplies (SIC 384), ophthalmic goods (SIC 385), photographic equipment and supplies (SIC 386), and watches, clocks, and watchcases (SIC 387). See Helen B. Munzer and Eugene Doody, *High Technology Employment: Massachusetts and Selected States 1975–1979* (Boston: Division of Employment Security, Job Market Research March 1981).

7. Massachusetts Division of Employment Security, ES202 Survey data.

8. Data further disaggregated to the three-digit industry level for the Lowell area are not available.

9. Massachusetts Division of Employment Security, *Annual Planning Report, Fiscal Year 1980, Lowell LMA* (Boston: Division of Employment Security) p. 41.

10. Bureau of Labor Statistics (BLS), U.S. Department of Labor, *Employment and Earnings, States and Regions, 1939–1982* (Washington, D.C.: U.S. Government Printing Office, 1983), and BLS, U.S. Department of Labor, *Supplement to Employment and Earnings* (Washington, D.C.: U.S. Government Printing Office, July 1983.)

11. See Edward J. Malecki, "High Technology and Local Economic Development," *Journal of the American Planning Association* (Summer 1984): 262–269; Edward J. Malecki, "Corporate Organization of R & D and the Location of Technological Activities," *Regional Science* 14 (1980): 219– 234; Allan R. Pred, *City-Systems in Advanced Economies* (New York: Wiley, 1977); John Rees and Howard Stafford, "High Technology Location and Regional Development: The Theoretical Base," in Office of Technology Assessment, *Technology, Innovation and Regional Economic Development* (Washington, D.C.: U.S. Government Printing Office, July 1984), Appendix A; Morgan D. Thomas, "Regional Economic Development and the Role of Innovation and Technological Change," in A.T. Thwaites and R.P. Oakey, eds., *The Regional Economic Impact of Technological Change* (New York: St. Martin's Press, 1985), pp. 13–35; Robert Premus, *Location of High Technology Firms and Regional Economic Development* (Washington, D.C.: Joint Economic Committee of Congress, June 1982).

12. The city of Lowell greatly extended the use of its UDAG funds by providing firms with lower-interest loans rather than outright grants, as was the tradition nationwide. As the loans were repaid, the funds were lent to other firms for industrial expansion.

13. Former U.S. Senator Paul Tsongas (D–Mass.), for example, was an extremely effective force behind initiating and coordinating the local redevelopment drive and in obtaining monies from both the public and the private sectors to support these efforts. Senator Tsongas, a Lowell native, proposed creation of the Lowell Development Finance Corporation (LDFC) and was instrumental in obtaining commitments from each of the city's banks. He also orchestrated the development and funding of the Lowell National Historical Park through Congress. In addition, in conjunction with City Manager Joseph Tully, he formulated the Lowell Plan.

   Dr. An Wang, founder and chairman of the board of Wang Laboratories, was another strong leader in Lowell's economic turnaround. Dr. Wang's impact on the Lowell area spreads far beyond the walls of Wang Laboratories. The Wang Institute, which awards master's degrees in computer engineering and software design, for example, was established with over $5 million from Dr. Wang. Operating independently of Wang Labs, this institution provides training for experienced workers from a variety of high-technology companies. In addition, the Wang Training Center, in which Wang employees from all over the world are trained, has been built in downtown Lowell. Dr. Wang is also a generous philanthropist who frequently contributes to local organizations.

   Many other individuals—including City Manager Joseph Tully, coauthor of the Lowell Plan; ex-city Manager William Taupier, credited with the idea of lending rather than granting outright the UDAG monies to firms; and Arthur Robbins, of Hilton Hotels, who built a first-class hotel in downtown Lowell—provided the personal impetus behind the Lowell redevelopment story. Along with the local media, these people helped generate widespread support from the "people of Lowell" for the high-technology-based economic revival. See Alan R. Earls, "Lowell Has a Comeback," *Mass High Tech*, 27 May–9 June 1985, Real Estate Section.

14. Defense contracts awarded to large high-technology firms in the area proved especially frustrating to small subcontracting firms. Whereas new contracts brought increased demands for the components that subcontractors produced, these firms often lost some of their trained employees to the larger firms in the process. One of these small employers noted that the large firms "just increase wages beyond our means to fill increased demand—mind you, their wages were above ours to begin with."

15. The International Institute of Lowell, a local organization that facilitates the entry and settling of new immigrants, helps these workers apply for jobs. Because many of these new arrivals speak little, if any, English, arrangements are sometimes made with employers for members of the institute staff to teach English to them during the first hour of each work

day. Such lessons, which take place at the factory, often begin with explanations of the plant's safety regulations and operating instructions for the machinery that the workers will be using.

16. Some firms in the traditional industries that remained in the Lowell area found a market "niche." A shoe firm, for example, produced women's white, dyable evening pumps; one textile plant specialized in tie labels; and another focused on automobile upholstery. See Michael E. Porter, *Competitive Strategy* (New York: Free Press, 1980) for discussion of the "niche" strategy for competitive advantage.

   In addition, some of the remaining textile firms use highly sophisticated, electronically controlled machinery. In contrast to the stereotypical image of textile mills, these factories were bright, quiet, and spotless, and required more highly skilled workers than in the past.

17. In one such attempt to establish a training program for stitchers, a school titled the course "Fashion Design" in order to attract students. Of the twenty-two students who enrolled, however, only three completed the program. Moreover, these three remained in their stitching jobs for less than three months, noting that it was not what they had expected.

Chapter 6: Education and Changing Skill Needs

1. This chapter draws heavily upon Patricia M. Flynn, "Technological Change, the 'Training Cycle,' and Economic Development," in John Rees, ed., *Technology, Regions and Policy* (Totowa, N.J.: Rowman & Littlefield, 1986): 282–308, and Patricia M. Flynn, "Production Life Cycles and Their Implications for Education and Training, Final Report, Grant No. NIE-82-0033-16 (Washington, D.C.: National Institute of Education, February 1984).

2. Donna Olszewski Shea, "Career Paths and Vocational Education," in Bruce Vermeulen and Peter B. Doeringer, eds., *Jobs and Training in the 1980s* (Boston: Martinus Nijhoff Publishing, 1981), pp. 88–117; Patricia Pannell, "Do Schools Adjust to Changing Job Demands? The Worcester Experience," *Thrust: The Journal for Employment and Training Professionals* 2, no. 1 (Winter/Spring 1980): 105–123; Paul Osterman, "The Structure of the Labor Market for Young Men," in Michael Piore, ed., *Institutional and Structural Views of Unemployment and Inflation* (New York: M.E. Sharpe, 1979), pp. 186–196.

3. Throughout this chapter the supply of occupationally trained graduates in the Lowell Labor Market Area is adjusted to account for the fact that the service areas of two of the regional vocational schools and the two community colleges are not synonymous with the Lowell labor market area. See Flynn, "Production Life Cycles," chap. 3.

4. Based on data from Massachusetts Division of Employment Security, *Manpower Requirements for Massachusetts by Occupation, by Industry 1970–1976* (Boston: Division of Employment Security, June 1973). There were no local labor market projections in the late 1960s and early to mid-1970s when these occupational training programs were designed and implemented. Occupational projection for the Lowell LMA 1974–1985 became available in the *Annual Planning Information FY1981* — too late to have been used for educational planning purposes during the period studied.

5. Two programs, accounting for less than 1 percent of all occupationally trained graduates, were introduced but subsequently phased out between 1970 and 1982. In the first instance, a training need was identified in health administration — an area in which employment had grown well above average in the 1960s. Occupational projections suggested a continuation of this trend. The training program devised at the associate's degree level, however, was found to be inappropriate, in that employers were requiring a minimum of a four-year degree program. Placement problems resulted in termination of the program.

   In the second case, although demand had neither grown relatively rapidly nor was projected to do so, a commercial photography program was introduced to fill an apparent training gap that existed in the local labor market. Realizing the limited market, the program was initiated on a small scale. Even though few in number, these graduates had trouble finding employment. More specifically, it appears that employers prefer negotiating contracts with experienced, self-employed photographers by the assignment, instead of hiring a full-time, in-house photographer. While graduates from this secondary-level program might eventually meet these criteria, immediate full-time placement was unlikely and the program was quickly phased out.

6. Clearly, graduates from many of these programs find jobs outside the high-technology industries. Secretaries and accountants, for example, have jobs in many industries. Furthermore, some high-technology workers, such as assemblers, production inspectors, and machine operatives, may be hired from a wide range of training programs. Hence, while changes and trends in these training programs are important for analyzing skill needs and educational responsiveness in various high-technology labor markets, specific occupational surpluses and shortages cannot be calculated by comparing the number of graduates with the actual number of jobs in these occupations.

7. Under the Vocational Education Act, as amended, Public Law 94-482. These data do not take into account monies awarded but cancelled in fiscal year 1982. The data are not available to determine the number of individuals who were subsequently denied training in these particular high-technology areas.

8. See, for example, Paula Goldman Leventman, *Professionals Out of Work* (New York: The Free Press, 1981); Berkeley Rice, "Down and Out along

Route 128," *New York Times Magazine* 1 November 1970, pp. 28–29, 93–98; "Technological Unemployment Again," *Bay State Business World*, January 22, 1975, pp. 1–2.

9. See, for example, Wellford Wilms, *Public and Proprietary Vocational Training: A Study of Effectiveness* (Lexington, Mass.: Lexington Books, 1975); Pannell, "Do Schools Adjust?"

10. These figures do not include Northern Essex Community College, from which the distribution of graduates by sex was not available.

11. Exclusive of the less formalized high-technology education programs discussed earlier in this chapter, for which comparable data were not available.

12. More detailed analysis of the responsiveness of higher education in New England to changing labor market conditions are available in several recent studies. These include, Marvin A. Sirbu, Jr., *High Technology Manpower in Massachusetts* (Cambridge: Center for Policy Alternatives, MIT, 1979); Peter B. Doeringer, Patricia Flynn, and Pankaj Tandon, "Market Influences on Higher Education: A Perspective for the 1980s," in John C. Hoy and Melvin H. Bernstein, eds., *Business and Academia: Partners in New England's Economic Renewal* (Hanover, N.H.: University Press of New England, 1981), pp. 26–74; Peter B. Doeringer and Patricia Flynn, "Manpower Strategies for Growth and Diversity in New England's High Technology Sector," in John C. Hoy and Melvin H. Bernstein, *New England's Vital Resource: The Labor Force* (Washington, D.C.: American Council on Education, 1982), pp. 11–35.

13. Bernard Wysocki, Jr., "More Firms Avoid Hiring MBA's Due to High Pay, Other Problems," *Wall Street Journal*, 17 September, 1980, p. 23.

14. Numerous studies have documented the tendency toward regular cycles of supply and demand imbalances in engineering fields because of sudden shifts in demand, long training times, and the lack of staffing alternatives. See, for instance, Richard B. Freeman and John A. Hansen, "Forecasting the Changing Market for College-Trained Workers," in Robert E. Taylor, Howard Rosen, and Frank C. Pratzner, eds., *Responsiveness of Training Institutions to Changing Labor Market Demands* (Columbus, Oh.: The National Center for Research in Vocational Education, Ohio State University, 1983), pp. 79–99; Richard B. Freeman, *The Market for College-Trained Manpower* (Cambridge: Harvard University Press, 1971); Walter Fogel and Daniel Mitchell, "Higher Education Decision-making and the Labor Market," in Margaret S. Gordon, ed., *Higher Education and the Labor Market* (New York: McGraw-Hill, 1974); Bruce Wilkinson, "Present Values of Lifetime Earnings for Different Occupations," *Journal of Political Economy* 74, no. 6 (1966): 556–572; Allan Cartter, *Ph.D's and the Academic Labor Market* (New York: McGraw-Hill, 1976); Kenneth Arrow and William Capron, "Dynamic Shortages and Price Rises: The Engineer-Scientist Case," *Quarterly Journal of Economics* 73 (May 1959): 292–308.

15. See, for instance, National Science Board, *Today's Problems, Tomorrow's Crises* (Washington, D.C.: National Science Foundation, 1983); New England Board of Higher Education (NEBHE), *A Threat to Excellence: The*

Preliminary Report of the Commission on Higher Education and the Economy of New England (Boston: NEBHE, 1982); National Commission on Excellence in Education, "A Nation at Risk," (Washington, D.C.: National Commission on Excellence in Education, 1983); James Botkin, Dan Dimancescu and Ray Stata, *Global Stakes: The Future of High Technology in America* (Cambridge, Mass.: Ballinger, 1982).

16. National Science Board, *Today's Problems*; also see Elizabeth L. Useem, *Low Tech Education in a High Tech World* (New York: The Free Press, 1986) for an in-depth look at the situation in high schools in cities and towns along Route 128.

Chapter 7: Technological Change and Human Resource Planning

1. Experimentation with new products or services, relatively high risks and R&D costs, small production volume and product variability, and demands for relatively highly skilled workers generally characterize industries in their initial stage of development. Emerging industries are relatively concentrated geographically. Agglomeration economies with respect to R&D, skilled labor pools, and support services such as marketing and communications are especially important in this early phase, particularly for small firms that lack the resources and internal capabilities of providing these factors. See Michael E. Porter, *Competitive Advantage* (New York: Free Press, 1980): chap. 8; Roy Rothwell and Walter Zegveld, *Innovation and the Small and Medium Sized Firm* (Boston: Kluwer-Nijhoff Publishing, 1982; Ray Oakey, *High Technology Small Firms* (New York: St. Martin's Press, 1984).

2. Benjamin Chinitz, "Contrasts in Agglomeration: New York and Pittsburgh," *American Economics Association, Papers and Proceedings* 50, no. 3 (1960): 279–289; Edward J. Malecki, "High Technology and Local Economic Development," *Journal of the American Planning Association* 50, no. 3 (Summer 1984): 262–269; Barry Bluestone and Bennett Harrison, *The Deindustrialization of America* (New York: Basic Books, 1982); Robert B. McKersie and Werner Sengenberger, *Job Losses in Major Industries* (Paris: OECD, 1983); M.E. Conroy, "Alternative Strategies for Regional Industrial Diversification," *Journal of Regional Science* 14, no. 1 (1974): 31–46; William L. Berentsen, "Regional Policy and Industrial Overspecialization in Lagging Regions," *Growth and Change* 9, no. 3 (1978): 9–13.

3. Massachusetts Division of Employment Security, "High Technology's Impact on the Massachusetts Economy since 1976," (Boston: Division of Employment Security, November 1985); Laura van Dam, "128's Layoff Season Threatens the Morale of Remaining Workers," *New England Business* 7, no. 12 (July 1, 1985): 33–35; "Hard Times in High Tech," *Newsweek*, April 22, 1985, pp. 50–51; Randell Smith, "IBM Sets High Goals, but Doubts Persist," *Wall Street Journal*, June 12, 1985, p. 6; Michael W.

Miller, "Slump in High Tech Casts a Long Shadow in the Silicon Valley," *Wall Street Journal*, July 24, 1985, p. 1.

4. Michael J. Piore, "On-the-Job Training and Adjustments to Technological Change," *Journal of Human Resources* 3, no. 4 (Fall 1968): 435–449; U.S. Congress, Office of Technology Assessment (OTA), *Technology and Structural Unemployment: Reemploying Displaced Adults* (Washington, D.C.: U.S. Government Printing Office, 1986), pp. 321–371; Eileen L. Collins and Lucretia Dewey Tanner, *American Jobs and the Changing Industrial Base* (Cambridge, Mass.: Ballinger, 1984).

5. Richard B. Freeman and John A. Hansen, "Forecasting the Changing Market for College-Trained Workers," in Robert E. Taylor, Howard Rosen, and Frank C. Pratzner, eds., *Responsiveness of Training Institutions to Changing Labor Market Demands* (Columbus, Oh.: The National Center for Research in Vocational Education, Ohio State University, 1983), pp. 79–99; David W. Stevens, *Employment Projections for Planning Vocational Technical Education Curricula: Mission Impossible?* (Columbia: University of Missouri, 1976); Harold Goldstein, "The Accuracy and Utilization of Occupational Forecasts," in Taylor, Rosen and Pratzner, eds., *Responsiveness of Training Institutions*, pp. 39–70; Patricia M. Flynn, "Identifying and Interpreting Skill Needs, A Case Study of Missouri," report prepared for Vocational Education Trends and Priorities: A Study of Vocational Education in Missouri, University of Missouri-Columbia (March 1984). (Mimeo).

6. U.S. Congress, Office of Technology Assessment (OTA), *Technology, Innovation and Regional Development, Census of the State Government Initiatives for High Technology Industrial Development* (Washington, D.C.: U.S. Government Printing Office, 1983).

7. See Patricia M. Flynn, "Vocational Education Policy and Economic Development: Balancing Short-Term and Long-Term Needs," in National Assessment of Vocational Education, *Design Papers for the National Assessment of Vocational Education* (Washington, D.C.: U.S. Department of Education, 1987): III-2–III-29.

8. Norton W. Grubb, "The Bandwagon Once More: Vocational Preparation for High Technology Occupations," *Harvard Educational Review* 54, no. 4 (November 1984): 429–451. Also see Angelo C. Gilli, Sr., "Vocational Education and High-Technology," *Journal of Studies in Technical Careers* 6, no. 3 (Summer 1984): 187–197; Sheldon Haber and Robert Goldfarb, "Labor Market Responses for Computer Occupations," *Industrial Relations* 17, no. 1 (February 1978): 53–63.

9. Richard B. Freeman, *The Market for College-Trained Manpower* (Cambridge: Harvard University Press, 1971); Freeman and Hansen, "Forecasting the Changing Market," pp. 79–99; Walter Fogel and Daniel Mitchell, "Higher Education Decision-Making and the Labor Market," in Margaret S. Gordon, ed., *Higher Education and the Labor Market* (New York: McGraw-Hill, 1974), pp. 453–502; Peter B. Doeringer and Patricia Flynn, "Manpower Strategies for Growth and Diversity in New England's High Technology Sector," in John C. Hoy and Melvin H. Bernstein, eds., *New England's Vital*

Resource: The Labor Force (Washington, D.C.: American Council on Education, 1982) pp. 11–35; Haber and Goldfarb, "Labor Market Responses."

10. Wellford W. Wilms, "The Nonsystem of Education and Training," in Peter B. Doeringer and Bruce Vermeulen, eds., *Jobs and Training in the 1980s* (Boston: Martinus Nijhoff Publishing, 1981), pp. 19–49; Patricia M. Flynn, "Occupational Education and Training: Goals and Performance," in Doeringer and Vermeulen, *Jobs and Training*, pp. 50–71.

11. Candee Harris, "Establishing High Technology Enterprises in Metropolitan Areas," in Edward M. Bergman, ed., *Local Economies in Transition*, (Durham, N.C.: Duke University Press, 1986), pp. 165–184; OTA, *Technology, Innovation and Regional Economic Development*.

12. Research indicates distinct differences between the contributions to local economic development of branch plants of established firms and those of "home-grown" new firms. In particular, branch plants are likely to provide a larger number of jobs, at least in the short run, than are new firms, and hence tend to appear immediately successful. Jobs at branch plants, however, are more apt to involve relatively standardized production activities than are those at newly created firms indigenous to the area. Jobs at branch plants generally are more vulnerable than those at "home-grown" firms to the dispersion to lower-costs areas as product demand or competition intensifies. See Gunter Krumme and Roger Hayter, "Implications of Corporate Strategies and Product Cycle Adjustments for Regional Employment Changes," in Lyndhurst Collings and David F. Walker, eds., *Locational Dynamics of Manufacturing Activities* (New York: Wiley, 1975), pp. 325–356; Morgan D. Thomas, "Growth Pole Theory, Technological Change and Regional Economic Development," *Papers of the Regional Science Association* 34 (1975): 3–25; Edward M. Bergman and Harvey A. Goldstein, "Dynamics, Structural Change and Economic Development Paths," in Edward M. Bergman, ed., *Local Economies in Transition* (Durham, N.C.: Duke University Press, 1986), pp. 84–110; Edward J. Malecki, "High Technology Sectors and Local Economic Development," in Bergman, *Local Economics in Transition*, pp. 129–142; S. Oster, "Industrial Search for New Locations: An Empirical Analysis," *Review of Economics and Statistics* 61, no. 2 (1979): 288–292; H.D. Watts, *The Branch Plant Economy* (London: Longman Publishers, 1981).

13. See John Rees and Howard Stafford, "High Technology Location and Regional Development: The Theoretical Base," in Office of Technology Assessment, *Technology, Innovation and Regional Economic Development*, (Washington, D.C.: U.S. Government Printing Office, July 1984), Appendix A; Thomas, "Growth Pole Theory"; Morgan D. Thomas, Regional Economic Development and the Role of Innovation and Technological Change," in A.T. Thwaites and R.P. Oakey, eds., *The Regional Economic Impact of Technological Change* (New York: St. Martin's Press, 1985), pp. 13–35; Edward J. Malecki, "Technology and Regional Development: A Survey," *International Regional Science Review* 8, no. 2 (1983): 89–125.

14. Ray Oakey, *High Technology Small Firms*; Richard W. Riche, Daniel E.

Hecker, and John U. Burgan, "High Technology Today and Tomorrow: A Small Slice of the Employment Pie," *Monthly Labor Review* 106, no. 11 (November 1983): 50–58; Lynn E. Browne, "Can High Tech Save the Great Lake States?" *New England Economic Review* (November/December 1983): 19–33; Malecki, "High Technology Sectors"; Martin Carnoy, "High Technology and International Labour Markets," *International Labour Review* 124, no. 6 (November/December): 643-660.

15. John S. Hekman, "The Future of High Technology Industry in New England: A Case Study of Computers," *New England Economic Review* (January/February 1980): 5–17; Harris, "Establishing High Technology Enterprises"; 166; Catherine Armington, Candee Harris, and Marjorie Odle, "Formation and Growth in High Technology Businesses: A Regional Assessment," in OTA, *Technology Innovation*, Appendix B.

16. Research also indicates that attempts to create "science parks" to spur local economic development tend to fail more often than succeed. Moreover, successful science parks appear to be more the result, rather than the cause, of an economic boom. See Oakey, *High Technology and Small Firms*, pp. 152–154. Also see P.S. Johnson and D.G. Cathcart, "New Manufacturing Firms and Regional Development: Some Evidence From the Northern Region, *Regional Studies* 13, no. 3 (1979): 269–280; David J. Storey, *Enterpreneurship and the Small Firm* (London: Croom Helm, 1982).

17. Browne, "Can High Tech Save?"; William J. Abernathy, Kim B. Clark, and Alan M. Kantrow, *Industrial Renaissance* (New York: Basic Books, 1983); John Rees, Ronald Briggs, and Donald Hicks, "New Technology in the United States' Machinery Industry: Trends and Implications," in A.T. Thwaites and R.P. Oakey, eds., *The Regional Economic Impact of Technological Change* (New York: St. Martin's Press, 1985), pp. 164–194.

18. Patricia M. Flynn, "Lowell: A High Technology Success Story," *New England Economic Review* (September/October 1984): 39–49.

19. Bureau of Labor Statistics (BLS), U.S. Department of Labor, *Employment and Earnings, States and Regions, 1939–82* (Washington, D.C.: U.S. Government Printing Office, 1983); BLS, U.S. Department of Labor Supplement to *Employment and Earnings* (Washington, D.C.: U.S. Government Printing Office, July 1983).

20. Bureau of the Census, U.S. Department of Commerce, *Detailed Characteristics of the Population: Massachusetts; Michigan; Pennsylvania* (Washington, D.C.: U.S. Government Printing Office, 1980).

21. Ibid.

BIBLIOGRAPHY

Abernathy, William J.; Kim B. Clark; and Alan M. Kantrow. *Industrial Renaissance*. New York: Basic Books, 1983.

Abernathy, William J., and James M. Utterback. "Patterns of Industrial Innovation." In Robert R. Rothberg, ed., *Corporate Strategy and Product Innovation*, 2nd ed. pp. 428–436. New York: The Free Press, 1981.

_____. "A General Model." In W.J. Abernathy, *The Productivity Dilemma*, ch. 4, pp. 68–84. Baltimore: Johns Hopkins Press, 1978.

Adler, Paul. "Rethinking the Skill Requirements of New Technologies." Working Paper 9-784-076. Boston, Mass.: Harvard University Graduate School of Business Administration, 1983. (Mimeo.)

Albin, Peter S. "Job Design within Changing Patterns of Technological Development." In Eileen Collins and Lucretia Dewey Tanner, eds., *American Jobs and the Changing Industrial Base*, pp. 125–162. Cambridge, Mass.: Ballinger, 1984.

Alderfer, E.B., and H.E. Michl. *Economics of American Industry*. New York: McGraw-Hill, 1942.

"America Rushes to High Tech for Growth." *Business Week*, March 28, 1983, p. 85.

Armington, Catherine; Candee Harris; and Marjorie Odle. "Formation and Growth in High Technology Businesses: A Regional Assessment," Appendix B. In Office of Technology Assessment (OTA), *Technology, Innovation and Regional Economic Development*. Washington, D.C.: U.S. Government Printing Office, July 1984.

Arrow, Kenneth J. "The Economic Implications of Learning by Doing." *The Review of Economic Studies* 29, no. 1 (July 1962): 155–173.

Arrow, Kenneth, and William Capron. "Dynamic Shortages and Price Rises: The Engineer-Scientist Case." *Quarterly Journal of Economics* 73 (May 1959): 292–308.

Attewell, Paul, and James Rule. "Computing and Organizations: What We Know and What We Don't Know." *Communications of the ACM* 27, no. 12 (December 1984): 1184–1192.

Baker, Elizabeth. *Technology and Women's Work*. New York: Columbia University Press, 1964.

Baran, Barbara. "Office Automation and Women's Work: The Technological Transformation of the Insurance Industry." In M. Castells, ed., *High Technology, Space and Society*. Beverly Hills: Sage Publications, 1985.

Barocci, Thomas A., and Paul Cournoyer. "Make or Buy: Computer Professionals in a Demand-Driven Environment." Working Paper 1342. Sloan School of Management, MIT, September 1982.

Barocci, Thomas A., and Kirsten R. Wever. "Information Systems Careers and Human Resource Management." Working Paper 1482. Sloan School of Management, MIT, September 1983.

Barkin, Solomon, ed. *Technological Change and Manpower Planning*. Paris: Organization of Economic Cooperation and Development, (OECD) 1967.

Becker, Gary. *Human Capital*. New York: Columbia University Press, 1964.

Benson, Ian, and John Lloyd. *New Technology and Industrial Change*. London: Kogan Page, 1983.

Berentsen, William L. "Regional Policy and Industrial Overspecialization in Lagging Regions." *Growth and Change* 9, no. 3 (1978): 9–13.

Bergman, Edward M., and Harvey A. Goldstein. "Dynamics, Structural Change and Economic Development Paths." In Edward M. Bergman, ed., *Local Economies in Transition*, pp. 84–110. Durham, N.C.: Duke University Press, 1986.

Best, Ethel. "The Change from Manual to Dial Operation in the Telephone Industry." Bulletin No. 10. Washington, D.C.: U.S. Women's Bureau, 1933.

Blair, Larry M. "Worker Adjustment to Changing Technology: Techniques, Processes, and Policy Considerations." In Eileen L. Collins and Lucretia Dewey Tanner, eds., *American Jobs and the Changing Industrial Base*, pp. 207–252. Cambridge, Mass.: Ballinger, 1984.

Bluestone, Barry. "Industrial Dislocation and Its Implications for Public Policy." In Keven Hollenbeck, Frank C. Pratzner, and Howard Rosen, eds., *Displaced Workers: Implications for Educational and Training Institutions*, pp. 45–68. Columbus, Oh.: The National Center for Research in Vocational Education, 1984.

Bluestone, Barry,, and Bennett Harrison. *The Deindustrialization of America*. New York: Basic Books, 1982.

Botkin, James; Dan Dimancescu; and Ray Stata. *Global Stakes: The Future of High Technology in America*. Cambridge, Mass.: Ballinger, 1982.

Braverman, Harry. *Labor and Monopoly Capital: The Degradation of Work in the Twentieth Century*. New York: Monthly Review, 1974.

Bright, James R. *Automation and Management*. Boston: Harvard University Graduate School of Business Administration, 1958.

_____. "Does Automation Raise Skill Requirements?" *Harvard Business Review* 36, no. 4 (July–August 1958): 85–98.

Browne, Lynn E. "Can High Tech Save the Great Lake States?" *New England Economic Review* (November/December 1983): 19–33.

Browning, Jon E. *How to Select a Business Site*. New York: McGraw-Hill, 1980.

Buckingham, Walter. *Automation: Its Impact on Business and People.* New York: Harper and Brothers Publishers, 1961.

Burns, Arthur F. *Production Trends in the United States since 1870.* New York: National Bureau of Economic Research, 1934.

Cain, P. and D. Treiman. "The D.O.T. as a Source of Occupational Data." *American Sociological Review* 46, no. 3 (1981): 235–278.

Carnevale, Anthony, and Harold Goldstein. *Employee Training: Its Changing Role and an Analysis of New Data.* Washington, D.C.: American Society for Training and Development, 1983.

Carnoy, Martin. "High Technology and International Labour Markets." *International Labour Review* 124, no. 6 (November/December 1985): 643–660.

Cartter, Allan. *Ph.D.'s and the Academic Labor Market.* New York: McGraw-Hill, 1976.

Chinitz, Benjamin. "Contrasts in Agglomeration: New York and Pittsburgh." *American Economics Association, Papers and Proceedings* 50, no. 3 (1960): 279–289.

Clague, Ewan. "Effects of Technological Change on Occupational Employment." In Organization of Economic Cooperation and Development (OECD), *The Requirements of Automated Jobs.* Paris, 1965.

Clark, G.L. "The Employment Relation and Spatial Division of Labor: A Hypothesis." *Annals of the Association of American Geographers* 71 (1981): 412–424.

Collins, Eileen L., and Lucretia Dewey Tanner. *American Jobs and the Changing Industrial Base.* Cambridge, Mass.: Ballinger, 1984.

Conroy, Michael E. "Alternative Strategies for Regional Industrial Diversification." *Journal of Regional Science* 14, no. 1 (1974): 31–46.

Cooper, Arnold C., and Daniel Schendel. "Strategic Responses to Technological Threats." *Business Horizons,* 19, no. 1 (February 1976): 61–69.

Cournoyer, Paul. "Mobility of Information Systems Personnel: An Analysis of a Large Computer Firm's Experience." Ph.D. dissertation, MIT, 1983.

Crossman, Edward R.F.W., and Stephen Laner. *The Impact of Technical Change on Manpower Skill Demand: Case-Study Data and Policy Implications.* Berkeley: University of California, 1969.

Dean, Joel. "Pricing Policies for New Products." *Harvard Business Review* 28, no. 6 (November 1950): 45–53.

Dhalla, Nariman K., and Sonia Yuspeh. "Forget the Product Life Cycle Concept." *Harvard Business Review* 54, no. 1 (January/February 1976): 102–112.

Division of Employment Security in cooperation with the U.S. Department of Labor. *Manpower Requirements for Massachusetts by Occupation, by Industry 1970–1976.* Boston, Mass.: June 1973.

Doeringer, Peter B. *Workplace Perspectives on Education and Training.* Boston: Martinus Nijhoff Publishing, 1981.

Doeringer, Peter B., and Patricia Flynn. "Manpower Strategies for Growth and Diversity in New England's High Technology Sector." In John C. Hoy and Melvin H. Bernstein, eds., *New England's Vital Resource: The Labor Force,* pp. 11–35. Washington, D.C.: American Council on Education, 1982.

Doeringer, Peter B.; Patricia Flynn; and Pankaj Tandon. "Market Influences on Higher Education: A Perspective for the 1980s." In John C. Hoy and Melvin H. Bernstein, eds., *Business and Academia: Partners in New England's Economic Renewal*, pp. 26–74. Hanover, N.H.: University Press of New England, 1981.

Doeringer, Peter B., and Michael J. Piore. *Internal Labor Markets and Manpower Analysis*. Armonk, N.Y.: M.E. Sharpe, 1985.

Doeringer, Peter B., David G. Terkla; and Gregory Topakian. *Invisible Factors and Local Economic Development*. New York: Oxford University Press, forthcoming.

Dublin, Thomas. *Women at Work: The Transformation of Work and Communities in Lowell, Massachusetts 1826–1860*. New York: Columbia University Press, 1979.

Earls, Alan R. "Lowell Has a Comeback." *Mass High Tech*, 27 May–9 June 1985, Real Estate Section.

Educational Research Services. *Education for a High Technology Future: The Debate over the Best Curriculum*. Arlington, Va., May 1983.

Eno, Arthur L., Jr. *Cotton Was King: A History of Lowell, Massachusetts*. Somersworth, N.H.: New Hampshire Publishing Company, 1976.

Erickson, Rodney A., and Thomas R. Leinbach. "Characteristics of Branch Plants Attracted to Nonmetropolitan Areas." In R.E. Lonsdale and H.L. Seyler, eds., *Nonmetropolitan Industrialization*, pp. 57–78. New York: Winston/Wiley, 1979.

Erivanksy, Arkadii. *A Soviet Automatic Plant*. Moscow: Foreign Language Publishing House, 1955.

Flamm, Kenneth. "Internationalization in the Semiconductor Industry." In Joseph Grunwald and Kenneth Flamm, eds., *The Global Factory*, pp. 38–136. Washington, D.C.: The Brookings Institution, 1985.

Flynn, Patricia M. "Employer Response to Skill Shortages: Implications for Small Business." *Proceedings of the Small Business Research Conference* Waltham, Mass.: Bentley College, 1981.

_____. "Occupational Education and Training: Goals and Performance." In Bruce Vermeulen and Peter B. Doeringer, eds., *Jobs and Training: Vocational Policy and the Labor Market*, pp. 50–71. Boston: Martinus Nijhoff Publishing, 1981.

_____. "Production Life Cycles and Their Implications for Education and Training." Final Report. Grant No. NIE-G-82-0033. Washington, D.C.: National Institute of Education, February 1984.

_____. "Identifying and Interpreting Skill Needs, A Case Study of Missouri." Report prepared for Vocational Education Trends and Priorities: A Study of Vocational Education in Missouri, University of Missouri, Columbia, March 1984. (Mimeo.)

_____. "Lowell: A High Technology Success Story." *New England Economic Review* (September/October 1984): 39–49.

_____. "The Impact of Technological Change on Jobs and Workers." Final Report. Grant No. 21-25-82-16. Washington, D.C.: U.S. Department of Labor, Employment and Training Administration, March 1985.

_____. "Technological Change, the 'Training Cycle' and Economic Development." In John Rees, ed., *Technology, Regions and Policy*, pp. 282–308. Totowa, N.J.: Rowman and Littlefield, 1986.

_____. "Vocational Education Policy and Economic Development: Balancing Short-Term and Long-Term Needs." In National Assessment of Vocational Education, *Design Papers for the National Assessment of Vocational Education* pp. III-2–III-29. Washington, D.C.: U.S. Department of Education, 1987.

Fogel, Walter, and Daniel Mitchell. "Higher Education Decision-making and the Labor Market." In Margaret S. Gordon, ed., *Higher Education and the Labor Market*, pp. 453–502. New York: McGraw-Hill, 1974.

Ford, David, and Chris Ryan. "Taking Technology to Market." *Harvard Business Review* 59, no. 2 (March/April 1981): 117–126.

Foster, Richard N. "A Call for Vision in Managing Technology." *Business Week*, May 24, 1982, pp. 24, 26, 28, 33.

_____. "To Exploit New Technology, Know When to Junk the Old." *Wall Street Journal*, 2 May 1983, p. 22.

Franchak, Stephen J. "Factors Influencing Vocational Education Program Decisions." In Robert E. Taylor, Howard Rosen, and Frank C. Pratzner, eds., *Responsiveness of Training Institutions to Changing Labor Market Demands*, pp. 267–293. Columbus, Oh.: The National Center for Research in Vocational Education, Ohio State University, 1983.

Freeman, Christopher. *The Economics of Industrial Innovation*, 2d ed. Cambridge, Mass.: The MIT Press, 1982.

Freeman, Richard B. *The Market for College-Trained Manpower*. Cambridge: Harvard University Press, 1971.

Freeman, Richard B., and John A. Hansen. "Forecasting the Changing Market for College-Trained Workers." In Robert E. Taylor, Howard Rosen, and Frank C. Pratzner, eds., *Responsiveness of Training Institutions to Changing Labor Market Demands*, pp. 79–99. Columbus, Oh.: The National Center for Research in Vocational Education, Ohio State University, 1983.

Gaston, Frank J. "Labor Market Conditions and Employer Hiring Standards." *Industrial Relations* 11, no. 2 (May 1972): 272–278.

Gilli, Angelo C., Sr. "Vocational Education and High-Technology." *Journal of Studies in Technical Careers* 6, no. 3 (Summer 1984): 187–197.

Glasmeier, Amy K.; Peter Hall; and Ann R. Markusen. "Recent Evidence on High-Technology Industries' Spatial Tendencies: A Preliminary Investigation." Working Paper No. 417. Institute of Urban and Regional Development, University of California-Berkeley, October 1983. (Mimeo.)

Goldstein, Harold. "The Accuracy and Utilization of Occupational Forecasts." In Robert E. Taylor, Howard Rosen, and Frank C. Pratzner, eds., *Responsiveness of Training Institutions to Changing Labor Market Demands*, pp. 39–70. Columbus, Oh.: The National Center for Research in Vocational Education, Ohio State University, 1983.

Goldstein, Harold, and Bryna Shore Fraser. "Training for Work in the Computer Age: How Workers Who Use Computers Get Their Training." Research Report Series, No. RR-85-09, National Commission on Employment

Policy, June 1985.

Gorham, William. "Trade Winds and Urban Industrial Change." Speech delivered at the International Symposium on Revitalization of Local Communities and the Role of Cities, Amagasaki, Japan, October 6, 1986 (available from The Urban Institute, Washington, D.C.).

Greenbaum, Joan. *In the Name of Efficiency: A Study of Change in Data Processing Work.* Philadelphia: Temple University Press, 1979.

Greene, Richard; Paul Harrington; and Robert Vinson. "High Technology Industry: Identifying and Tracking an Emerging Source of Employment Strength." *New England Journal of Employment and Training* (Fall 1983).

Greenhalgh, N.H. "Automation at Gawith's Bakery." *Personnel Practice Bulletin* 18, no. 2 (June 1962): 28.

Grubb, Norton. W. "The Bandwagon Once More: Vocational Preparation for High Technology Occupations." *Harvard Educational Review* 54, no. 4 (November 1984): 429–451.

Gruenstein, John M.L. "Targeting High Technology in the Delaware Valley." *Business Review* (May/June 1984): 3–14.

Haber, Sheldon, and Robert Goldfarb. "Labor Market Responses for Computer Occupations." *Industrial Relations* 17, no. 1 (February 1978): 53–63.

Haddy, Pamela. "Some Thoughts on Automation in a British Office." *Journal of Industrial Economics* 6 (1958): 161–170.

Hall, P.; A. Markusen; R. Osborn; and B. Wachsman. "The American Computer Software Industry: Economic Development Prospects." *Built Environment* 9, no. 1 (1983): 29–39.

"Hard Times in High Tech." *Newsweek*, April 22, 1985, pp. 50–51.

Harrington, James W., Jr. "Learning and Locational Change in the American Semiconductor Industry." In John Rees, ed., *Technology, Regions and Policy*, pp. 120–137. Totowa, N.J.: Rowman and Littlefield, 1986.

Harris, Candee. "Establishing High Technology Enterprises in Metropolitan Areas." In Edward M. Bergman, ed., *Local Economies in Transition*, pp. 165–184. Durham, N.C.: Duke University Press, 1986.

Harrison, Bennett. "Rationalization, Restructuring and Industrial Reorganization of Older Regions: The Economic Transformation of New England since World War II." Working Paper No. 72. Joint Center for Urban Studies of M.I.T. and Harvard University, Cambridge, Mass., February 1982. (Mimeo.)

Hayes, Robert H., and Steven C. Wheelwright. "Link Manufacturing Process and Product Life Cycles." *Harvard Business Review* 57, no. 1 (January/February 1979): 133–140.

———. "The Dynamics of Process-Product Life Cycles." *Harvard Business Review* 57, no. 2 (March/April 1979): 127–136.

Hekman, John. "The Future of High Technology Industry in New England: A Case Study of Computers." *New England Economic Review* (January/February 1980): 5–17.

———. "Can New England Hold onto Its High Technology Industry?" *New England Economic Review* (March/April 1980): 35–44.

_____. "The Product Cycle and New England Textiles." *Quarterly Journal of Economics* 94, no. 4 (June 1980): 697–717.

"High Technology Makes Lowell a Model of Reindustrialization." *New York Times*, 10 August 1982, p. 1.

Hirsch, Seev. *Location of Industry and International Competitiveness*. Oxford, England: The Clarendon Press, 1967.

Hirsch, Seev. "The United States Electronics Industry in International Trade." In Louis T. Wells, Jr., ed., *The Product Life Cycle and International Trade*, pp. 39–52. Cambridge: Harvard University Press, 1972.

Hollenbeck, Kevin; Frank C. Pratzner; and Howard Rosen. *Displaced Workers: Implications for Educational and Training Institutions*. Columbus, Oh.: The National Center for Research in Vocational Education, 1984.

Hoover, Edgar M. *The Location of Economic Activity*. New York: McGraw-Hill, 1948.

Horowitz, Morris, and Irwin Herrnstadt. "Changes in Skill Requirements of Occupations in Selected Industries." In *Report of the National Commission on Technology, Automation, and Economic Progress*. Washington, D.C.: U.S. Government Printing Office, 1966.

Howard, Robert. "Second Class in Silicon Valley." *Working Papers* (September/October 1981): 21–31.

Jaffe, Abram J., and Joseph Froomkin. *Technology and Jobs: Automation in Perspective*. New York: Praeger, 1966.

Johnson, P.S., and D.G. Cathcart. "New Manufacturing Firms and Regional Development: Some Evidence From the Northern Region." *Regional Studies* 13, no. 3 (1979): 269–280.

Kamien, Morton I., and Nancy L. Schwartz. *Market Structure and Innovation*. Cambridge, England: Cambridge University Press, 1982.

Kanter, Rosabeth Moss. "Variations in Managerial Career Structures in High-Technology Firms: The Impact of Organizational Characteristics on Internal Labor Market Patterns." In Paul Osterman, ed., *Internal Labor Markets*, pp. 109–132. Cambridge, Mass.: The MIT Press, 1984.

Karmin, Monroe W. "High Tech: Blessing or Curse?" *U.S. News and World Report* 96, no. 2 (January 16, 1984): 38–44.

Kennegott, George F. *The Record of a City: A Social Survey of Lowell, Massachusetts*. New York: MacMillan, 1912.

Kiechell, Walter, III. "The Decline of the Experience Curve." *Fortune*, October 5, 1981, pp. 139–140, 144, 146.

Killingsworth, Charles. "The Automation Story: Machines, Manpower and Jobs." In C. Markhan, *Jobs, Men and Machines*, pp. 15–87. New York: Praeger, 1964.

Koch, D.L.; W.N. Cox; D.W. Steinhauser; and P.V. Whigham. "High Technology: The Southeast Reaches out for Growth Industry." *Economic Review* 68, no. 9 (September 1983): 4–19.

Kraft, Philip. *Programmers and Managers: The Routinization of Programming in the United States*. New York: Springer-Verlag, 1977.

Krumme, Gunter, and Roger Hayter. "Implications of Corporate Strategies and Product Cycle Adjustments for Regional Employment Changes." In Lyndhurst Collins and David Walker, eds., *Locational Dynamics of Manufacturing Activities*, pp. 325–356. New York: Wiley, 1975.

Kuhn, Sarah. *Computer Manufacturing in New England*. Cambridge, Mass.: Joint Center for Urban Studies of MIT and Harvard University, April 1982.

Kuznets, Simon. *Secular Movement in Production and Prices*. Boston: Houghton Mifflin, 1930.

Lester, Richard. *Hiring Practices and Labor Competition*. Princeton: Industrial Relations Section, Princeton University, 1954.

Leventman, Paula Goldman. *Professionals Out of Work*. New York: The Free Press, 1981.

Levin, Henry, and Russell Rumberger. *The Educational Implications of High Technology*. Palo Alto, Calif.: Institute for Research on Educational Finance and Governance, Stanford University, 1983.

_____. "The Low-Skill Future of High Tech." *Technology Review* 86, no. 6 (August/September 1983): 18–21.

Levitan, Sar, and Harold Sheppard, "Technological Change and the Community." In G. Somers, E. Cushman, and N. Weinberg, *Adjusting to Technological Change*. New York: Harper and Row, 1963.

Levitt, Theodore, "Exploit the Product Life Cycle." *Harvard Business Review* 43, no. 6 (November/December 1965): 81–94.

Lipstrev, Otis, and Kenneth A. Reed. *Transition to Automation: A Study of People, Production and Change*. Boulder: University of Colorado Press, 1964.

"Lowell: From Riches to Rags and Back Again." *Dun's Review*, July 1980, 38–39.

Lusterman, Seymour. *Education in Industry*. New York: The Conference Board, 1977.

Malecki, Edward J. "Firm Size, Location and Industrial R & D: A Disaggregated Analysis." *Review of Business and Economic Research* 16, no. 2 (1980): 29–42.

_____. "Corporate Organization of R & D and the Location of Technological Activities." *Regional Studies* 14 (1980): 219–234.

_____. "Technology and Regional Development: A Survey." *International Regional Science Review* 8, no. 2 (1983): 89–125.

_____. "High Technology and Local Economic Development" *Journal of the American Planning Association* 50, no. 3 (Summer 1984): 262–269.

_____. "High Technology Sectors and Local Economic Development." In Edward M. Bergman, ed., *Local Economies in Transition*, pp. 129–142. Durham, N.C.: Duke University Press, 1986.

Malm, F. Theodore. "Recruiting and the Functioning of Labor Markets." *Industrial Labor Relations Research* 7, no. 4 (July 1954): 507–525.

Mansfield, Edwin. *The Economics of Technological Change*. New York: W.W. Norton, 1968.

_____. *Industrial Research and Technological Innovation*. New York: W.W. Norton and Company, 1968.

Mansfield, E.; J. Rapoport; A. Romeo; E. Villani; S. Wagner; and F. Husic. *The Production and Application of New Industrial Technology*. New York: W.W.

Norton and Company, 1977.

Markusen, Ann R. "High Tech Jobs, Markets and Economic Development Prospects: Evidence from California." In Peter Hall and Ann R. Markusen, eds., *Silicon Landscapes*, pp. 35–48. Boston: Allen and Unwin, 1985.

———. *Profit Cycles, Oligopoly and Regional Development*. Cambridge, Mass.: The MIT Press, 1985.

———. "Defense Spending and the Geography of High-Tech Industries." In John Rees, eds., *Technology, Regions and Policy*, pp. 94–119. Totowa, N.J.: Rowman and Littlefield, 1986.

Maryland Department of Employment Security. *The Impact of Technological Change in the Banking Industry*. Baltimore: Automation Manpower Services Program, June 1967.

Massachusetts Division of Employment Security. *Annual Planning Report, Fiscal Year 1980, Lowell LMA* (Boston: Division of Employment Security) p. 41.

———. *High Technology's Impact on the Massachusetts Economy Since 1976*. Boston: Division of Employment Security, November 1985.

Massey, Doreen. "In What Sense a Regional Problem?" *Regional Studies* 13, no. 2 (1979): 231–241.

Mathias, E.P. "Management of Technological Change in Railways." In C.P. Thakur and G.S. Aurora, eds., *Technological Change and Industry*, pp. 101–133. New Delhi: Shri Ram Centre for Industrial Relations, 1971.

Matthaei, Julie A. *An Economic History of Women in America*. New York: Schocken Books, 1982.

McKersie, Robert B., and Werner Sengenberger. *Job Losses in Major Industries*. Paris: Organization of Economic Cooperation and Development (OECD), 1983.

Meserve, H.C. *Lowell: An Industrial Dream Come True*. Boston: National Association of Cotton Manufacturers, 1923.

Menzies, Heather. *Women and the Chip: Case Studies of the Effects of Informatics on Employment in Canada*. Montreal: Institute for Research on Public Policy, 1981.

Miller, Michael W. "Slump in High Tech Casts a Long Shadow in the Silicon Valley." *Wall Street Journal*, 24 July 1985, p. 1.

Mitchell, Sir Stuart. "Planning Coordination: British Railways." In Organization of Economic Cooperation and Development (OECD), *Adjustment of Workers to Technological Change at the Plant Level*, Supplement to the Final Report, pp. 184–196. Paris, 1966.

Mumford, Enid, and Olive Banks. *The Computer and the Clerk*. London: Routledge and Kegan Paul, 1967.

Munzer, Helen, and John Doody. *High Technology Employment: Massachusetts and Selected States, 1975–1979*. Boston: Division of Employment Security, Job Market Research, March 1981.

Myers, Charles A., and George P. Schultz. *The Dynamics of a Labor Market: A Study of the Impact of Employment Changes on Labor Mobility, Job Satisfactions and Company and Union Policies*. New York: Prentice-Hall, 1951.

National Commission on Excellence in Education. *A Nation at Risk.*

Washington, D.C.: National Commission on Excellence in Education, 1983.

National Science Board. *Today's Problems, Tomorrow's Crises.* Washington, D.C.: National Science Foundation, 1983.

National Science Foundation (NSF). *The Process of Technological Innovation: Reviewing the Literature.* Washington, D.C.: NSF, May 1983.

Nelson, R.R. "Research on Productivity Growth and Productivity Differences: Dead Ends and New Departures." *Journal of Economic Literature* 19, no. 3, (1981): 1029–1064.

Nelson, R.R., and S.G. Winter. "Neoclassical vs. Evolutionary Theories of Economic Growth: Critique and Prospectus." *Economic Journal* 84 (1974): 886–905.

New England Board of Higher Education (NEBHE). *A Threat to Excellence: The Preliminary Report of the Commission on Higher Education and the Economy of New England.* Boston: NEBHE, 1982.

"No Shuttlecocks at Parlin." *Fortune*, February 1961, pp. 189–190.

Norton, R.D., and John Rees. "The Product Cycle and the Spatial Decentralization of American Manufacturing." *Regional Studies* 13, no. 2 (1979): 141–151.

Oakey, Ray. *High Technology Small Firms.* New York: St. Martin's Press, 1984.

Organization of Economic Cooperation and Development (OECD). *The Requirements of Automated Jobs.* Paris, 1965.

Osborn, David G. "Automatic Data Processing in the Large Company." In Howard Boone Jacobson and Joseph S. Roucek, eds., *Automation and Society.* New York: Philosophical Library, 1959.

Oster, Sharon. "Industrial Search for New Locations: An Empirical Analysis." *Review of Economics and Statistics* 61, no. 2 (1979): 288–292.

Osterman, Paul, ed. *Internal Labor Markets.* Cambridge, Mass.: The MIT Press, 1984.

Osterman, Paul. "The Structure of the Labor Market for Young Men." In Michael Piore, ed., *Institutional and Structural Views of Unemployment and Inflation*, pp. 186–196. New York: M.E. Sharpe, 1979.

Pannell, Patricia. "Do Schools Adjust to Changing Job Demands? The Worcester Experience." *Thrust: The Journal for Employment and Training Professionals* 2, no. 1 (Winter/Spring 1980): 105–123.

Parker, Margaret Terrell. *Lowell: A Study of Industrial Development.* New York: Kennikat, 1940.

Paul, B.R. "The Introduction of Electronic Data Processing in Life Assurance." *Personnel Practice Bulletin* 18, no. 2 (June 1962): 7–11.

Peitchinis, Stephen G. *The Effect of Technological Changes on Educational and Skill Requirements of Industry.* Ottawa: Department of Industry, Trade and Commerce, 1978.

———. *Computer Technology and Employment.* London: Macmillan, 1983.

Pennsylvania State Employment Service. *The Effect of Automation on Occupations and Workers in Pennsylvania.* May 1965.

Phalen, Clifton W. "Automation and the Bell System." In Howard B. Jacobson and Joseph S. Roucek, eds., *Automation and Society.* New York: Philosophical

Library, 1959.

Philips, Robyn Swaim, and Avis C. Vidal. "Restructuring and Growth Transitions of Metropolitan Economies." In Edward M. Bergman, ed., *Local Economies in Transition*. pp. 59–83. Durham, N.C.: Duke University Press, 1986.

Piore, Michael J. "On-the-Job Training and Adjustment to Technological Change." *Journal of Human Resources* 3, no. 4 (Fall 1968): 435–449.

Porter, Michael E. *Competitive Strategy*. New York: Free Press, 1980.

Pred, Allan R. *City-Systems in Advanced Economies*. New York: Wiley, 1977.

Premus, Robert. *Location of High Technology Firms and Regional Economic Development*. Washington, D.C.: Joint Economic Committee of Congress, June 1982.

Rees, John; Ronald Briggs; and Donald Hicks. "New Technology in the United States' Machinery Industry: Trends and Implications." In A.T. Thwaites and R.P. Oakey, eds., *The Regional Economic Impact of Technological Change*, pp. 164–194. New York: St. Martin's Press, 1985.

Rees, John, and Howard Stafford. "High Technology Location and Regional Development: The Theoretical Base." In Office of Technology Assessment (OTA), *Technology, Innovation and Regional Economic Development*, Appendix A. Washington, D.C.: U.S. Government Printing Office, July 1984.

Report of the National Commission on Technology, Automation and Economic Progress. Vol. 1. Washington, D.C.: U.S. Government Printing Office, 1966.

Reynolds, Lloyd G. *The Structure of Labor Markets: Wages and Labor Mobility in Theory and Practice*. New York: Harper and Brothers, 1951.

Rice, Berkeley. "Down and Out along Route 128." *New York Times Magazine*, 1 November 1970, 28–29, 93–98.

Riche, Richard W.; Daniel E. Hecker; and John U. Burgan. "High Technology Today and Tomorrow: A Small Slice of the Employment Pie." *Monthly Labor Review* (November 1983): 50–58.

Rogers, Everett, and Judith Laisen. *Silicon Valley Fever*. New York: Basic Books, 1984.

Rones, Philip L. "The Labor Market Problems of Older Workers." *Monthly Labor Review* 106, no. 5 (May 1983): 3–12.

Rosenberg, Nathan. *Perspectives on Technology*. Armonk, N.Y.: M.E. Sharpe, 1976.

_____. *Inside the Black Box: Technology and Economics*. Cambridge, England: Cambridge University Press, 1982.

Ross A., and J. Ross. "Employment Problems of Older Workers." *Studies in Unemployment*. Washington, D.C.: Senate Special Committee on Unemployment Problems, 1959.

Rothwell, Roy, and Walter Zegveld. "National Coal Board." In Roy Rothwell and Walter Zegveld, eds., *Technical Change and Employment*, pp. 66–76. New York: St. Martin's Press, 1979.

Rothwell, Roy, and Walter Zegveld. *Innovation and the Small and the Medium Sized Firm*. Boston: Kluwer Nijhoff Publishing, 1981.

_____. *Reindustrialization and Technology*. New York: M.E. Sharpe, 1985.

Rumberger, Russell. "Changing Skill Requirements of Jobs in the U.S. Economy." *Industrial Labor Relations Review* 34, no. 4 (1981): 578–590.

Salmans, Sandra. "Pilkington's Progressive Shift." *Management Today* 163 (September 1980): 66–73.

Saxenian, AnnaLee. "Silicon Chips and Spatial Structure: The Industrial Basis of Urbanization in Santa Clara County, California." Working Paper 345. Institute of Urban and Regional Development, University of California, March 1981.

Saxenian, AnnaLee. "The Urban Contractions of Silicon Valley: Regional Growth and Restructuring of the Semiconductor Industry." *International Journal of Urban and Regional Research* 7 (1983): 237–261.

Schultz, George, and Arnold Weber. *Strategies For the Displaced Worker.* New York: Harper and Row, 1966.

Schumpeter, Joseph. *Capitalism, Socialism and Democracy.* New York: Harper and Row, 1942.

Segal Quince Wickstead. *The Cambridge Phenomenon: The Growth of High Technology Industry in a University Town.* Cambridge, England: Segal Quince Wickstead, 1985.

Shanklin, William L., and John K. Ryans, Jr. *Marketing High Technology* Lexington, Mass.: Lexington Books, 1984.

Shea, Donna Olszewski. "Career Paths and Vocational Education." In Bruce Vermeulen and Peter B. Doeringer, eds., *Jobs and Training in the 1980s,* pp. 88–117. Boston: Martinus Nijhoff Publishing, 1981.

Shepard, Jon. *Automation and Alienation: A Study of Office and Factory Workers.* Cambridge: The MIT Press, 1971.

Sheppard, Harold L., and James L. Stern. "Impact of Automation on Workers in Supplier Plants." *Labor Law Journal* 8, no. 10 (1957): 714–718.

Sirbu, Marvin A., Jr. *High Technology Manpower in Massachusetts.* Cambridge: Center for Policy Alternatives, MIT, 1979.

Smith, Randell. "IBM Sets High Goals, but Doubts Persist." *Wall Street Journal,* 12 June 1985, 6.

Stanford Research Institute. *Management Decisions to Automate.* Adapted by Harry F. Bonfils. Washington, D.C.: U.S. Department of Labor, Office of Manpower, Automation and Training, 1965.

Stanton, Erwin S. *Successful Personnel Recruiting and Selection.* New York: AMACOM, 1977.

Steiber, Jack. "Manpower Adjustments to Automation and Technological Change in Western Europe." In *Report of the National Commission on Technology, Automation and Economic Progress.* Vol. 1. Washington, D.C.: U.S. Government Printing Office, 1966.

Stevens, Benjamin, and Carolyn Brackett. *Industrial Location: A Review and Annotated Bibliography of Theoretical, Empirical and Case Studies.* Philadelphia: Regional Science Research Institute, 1967.

Stevens, David W. *Employment Projections for Planning Vocational Technical Education Curricula: Mission Impossible?* Columbia: University of Missouri 1976.

Stobough, Robert. "The Neotechnology Account of International Trade: The Case of Petrochemicals." In Louis T. Wells, Jr., ed., *The Product Life Cycle and International Trade*, pp. 83–105. Cambridge, Mass.: Harvard University Press, 1972.

Storey, David J. *Entrepreneurship and the Small Firm*. London: Croom Helm, 1982.

Stromberg, Ann H., and Shirley Harkness, eds. *Women Working: Theories and Facts in Perspective*. Palo Alto, Calif.: Mayfield, 1978.

Summers, G.F.; S.D. Evans; F. Clemente; E.M. Beck; J. Minkoff; and E. Elwood. *Industrial Invasion of Non-Metropolitan America: A Quarter Century of Experience*. New York: Praeger, 1976.

"Technological Unemployment Again." *Bay State Business World*, January 22, 1975, pp. 1–2.

Thomas, Morgan D. "Growth Pole Theory, Technological Change, and Regional Economic Development." *Papers of the Regional Science Association* 34 (1975): 3–25.

_____. "Regional Economic Development and the Role of Innovation and Technological Change." In A.T. Thwaites and R.P. Oakey, eds., *The Regional Economic Impact of Technological Change*, pp. 13–35. New York: St. Martin's Press, 1985.

Thwaites, A.T., and Raymond P. Oakey, eds. *The Regional Economic Impact of Technological Change*. New York: St. Martin's Press, 1985.

Tilton, John E. *International Diffusion of Technology: The Case of Semiconductors*. Washington, D.C.: The Brookings Institution, 1971.

Tomaskovic-Derey, Donald, and S.M. Miller, "Can High-Tech Provide the Jobs?" *Challenge* 26, no. 2 (May/June 1983): 57–63.

U.S. Congress, Office of Technology Assessment (OTA). *Automation and the Workplace: Selected Labor, Education and Training Issues*. Washington, D.C.: U.S. Government Printing Office, 1983.

_____. *Technology, Innovation and Regional Economic Development: Census of the State Government Initiatives for High Technology Industrial Development*. Washington, D.C.: U.S. Government Printing Office, 1983.

_____. *International Competitiveness in Electronics*. Washington, D.C.: U.S. Government Printing Office, November 1983.

_____. *Computerized Manufacturing Automation, Employment, Education and the Workplace*. Washington, D.C.: U.S. Government Printing Office, April 1984.

_____. *Technology, Innovation and Regional Economic Development*. Washington, D.C.: U.S. Government Printing Office, July 1984.

_____. *Technology and Structural Unemployment: Reemploying Displaced Adults*. Washington, D.C.: U.S. Government Printing Office, 1986.

U.S. Department of Commerce. Bureau of the Census. *Detailed Characteristics of the Population: Massachusetts; Michigan; Pennsylvania*. Washington, D.C.: U.S. Government Printing Office, 1980.

U.S. Department of Labor. Bureau of Labor Statistics. *Employment and Earnings, States and Regions, 1939–1982*. Washington, D.C.: U.S. Government Printing Office, 1983.

_____. *Supplement to Employment and Earnings*. Washington, D.C.: U.S. Government Printing Office, July 1983.

Useem, Elizabeth L. *Low Tech Education in a High Tech World*, New York: The Free Press, 1986.

Utterback, James M., and William J. Abernathy. "A Dynamic Model of Process and Product Innovation." *Omega* 3, no. 6 (1975); 639–656.

van Auken, K.G., Jr. "Plant Level Adjustments to Technological Change." *Monthly Labor Review* 76, no. 4 (April 1953): 388–391.

_____. "Personal Adjustment to Technological Change." In Howard B. Jacobson and Joseph S. Roucek, eds., *Automation and Society*, pp. 387–391. New York: Philosophical Library, 1959.

van Dam, Laura. "128's Layoff Season Threatens the Morale of Remaining Workers." *New England Business* 7, no. 12, (July 1, 1985): 33–35.

van Duijn, J.J. *The Long Wave in Economic Life*. London: George Allen and Unwin, 1983.

Vermeulen, Bruce, and Susan Hudson-Wilson. "The Impact of Workplace Practices on Education and Training Policy." In Bruce Vermeulen and Peter Doeringer, eds., *Jobs and Training in the 1980s*, pp. 72–87. Boston: Martinus Nijhoff Publishing, 1981.

Vernon, Raymond. "International Investment and International Trade in the Product Cycle." *Quarterly Journal of Economics* 80, no. 2 (May 1966): 190–207.

_____, ed. *The Technology Factor in International Trade*. New York: Columbia University Press, 1970.

_____. "The Product Cycle Hypothesis in a New International Environment." *Oxford Bulletin of Economics and Statistics* 41, no. 4 (1979): 255–267.

Vinson, Robert, and Paul Harrington. *Defining High Technology Industries in Massachusetts*. Boston: Department of Manpower Development, 1979.

Walker, Charles R. *Toward the Automatic Factory: A Case Study of Men and Machines*. New Haven, Conn.: Yale University Press, 1957.

_____. "Changing Character of Human Work under the Impact of Technological Change." In *Report of the National Commission on Technology, Automation and Economic Progress*, Vol. 1. Washington, D.C.: U.S. Government Printing Office, 1966.

Walker, W.B. *Industrial Innovation and International Trading Performance*. Greenwich, Conn.: JAI Press, 1979.

Wasson, Chester R. *Dynamic Competitive Strategy and Product Life Cycles*, 3d ed. Austin: Austin Press, 1978.

Watts. H.D. *The Branch Plant Economy*. London: Longman Publishers, 1981.

Weber, Arnold. "The Interplant Transfer of Displaced Employees." In G. Somers, E. Cushman, and N. Weinberg, eds., *Adjusting to Technological Change*, pp. 95–143. New York: Harper and Row, 1963.

Weber, Arnold. "Variety in Adaptation to Technological Change." In Organization of Economic Cooperation and Development (OECD). *The Requirements of Automated Jobs*. Paris, 1965.

Weiss, M.A. "High Technology Industries and the Future of Employment." *Built Environment* 9, no. 1 (1983): 51–60.

Wells, Louis T., Jr. "International Trade: The Product Life Cycle Approach." In Louis T. Wells, Jr., ed. *The Product Life Cycle and International Trade*, pp. 3–33. Cambridge: Harvard University Press, 1972.

———, ed. *The Product Life Cycle and International Trade*. Cambridge, Mass.: Harvard University Press, 1972.

Wilkins, Russell. *Microelectronics and Employment in Public Administration: Three Ontario Municipalities, 1976–1980*. Ontario: Ministry of Labor, July 1981.

Wilkinson, Barry. "Managing with New Technology." *Management Today*, October 1982, 33–40.

Wilkinson, Bruce. "Present Values of Lifetime Earnings for Different Occupations." *Journal of Political Economy* 74, no. 6 (1966): 556–572.

Willener, Alfred. "A French Steelworks—Closure of an Old Rolling Mill: Problems of Co-ordinated Transfers of Personnel." In Organization of Economic Cooperation and Development (OECD), *Adjustment of Workers to Technological Change at the Plant Level*, Supplement to the Final Report, pp. 145–158. Paris, 1966.

Wilms, Wellford W. *Public and Proprietary Vocational Training: A Study of Effectiveness*. Lexington, Mass.: Lexington Books, 1975.

———. "The Nonsystem of Education and Training." In Peter B. Doeringer and Bruce Vermeulen, eds., *Jobs and Training in the 1980s*, pp. 19–49. Boston: Martinus Nijhoff Publishing, 1981.

Wisconsin State Employment Services. "A Large Insurance Company Automates: Workforce Implications of Computer Conversion." Automation Manpower Services Program Demonstration Project No. 3. Madison, April 1964.

Wysocki, Bernard, Jr. "More Firms Avoid Hiring MBA's due to High Pay, Other Problems." *Wall Street Journal*, 17 September 1980, p. 23.

ANNOTATED BIBLIOGRAPHY
FOR THE CASE-STUDY
DATA BASE

Banks, Olive. *The Attitudes of Steelworkers to Technical Change*. Liverpool, England: Liverpool University Press, 1960. Discussion of the attitudes of workers and the actual consequences of changes resulting from the introduction of a new steelworks in a large British steel company in the early 1950s.

Bannon, Ken, and Nelson Samp. "The Impact of Automation of Ford–U.A.W. Relationships." *Monthly Labor Review* 81, no. 6 (June 1958): 612–615. Study of how workers and management were affected by the construction and operation of a new, automated stamping plant (in Cleveland) of the Ford Motor Company in the mid-1950s.

Barkin, Solomon, ed. *Technological Change and Manpower Planning*. Paris: Organization of Economic Cooperation and Development, 1967.

"A Policy of Continuous Change with a Stable Staff," pp. 58–64. Eleven-year study (1953–1964) documenting the experiences of a relatively large French hosiery and knitwear company in which the technical machinery and production methods were continually being updated.

"Adjustments to Central Planning and Production Control in a New Shipyard," pp. 158–167. Study of the production and staffing of a modernized shipyard (Arendal) in Sweden from 1957 to 1963. The new shipyard was built to expand production beyond that of an older shipyard (Gotaverken).

"Changed Manpower Needs in Conversions to Dial Telephones," pp. 269–278. Study of three large regional telephone companies undergoing some of the last conversions within AT&T to modernized direct dialing and automatic message accounting in the United States in the 1950s and early 1960s.

"Concentration in a Nationalised Industry," pp. 50–57. Study of the French nationalized gas industry (Gaz de France) from 1946 to 1964. Focuses on the adjustment of displaced workers after the modernization of some plants and the closing of obsolete plants, which resulted in reduction of the gas company's employees by 50 percent (from 12,000 to 6,400) over the eighteen-year period.

"Concentration of Production and Transfer of Employees," pp. 129–137. Study of two Norwegian factories consolidated so that the company could take advantage of economies of scale in technical advances in the 1950s.

"Data Processing and Manpower Savings in Public Administration," pp. 92–101. The effects of computerizing the file data of the Pension Department of the West German federal government from 1958 to 1960.

"Distribution of Liquefied Gas — A Ten-Year Programme of Decentralization and Mechanisation," pp. 65–71. Study of the mechanization and decentralization of the bottling procedure in a French liquefied gas distributorship from 1958 to 1960.

"Flexibility and Control in Automation and Telephone Traffic," pp. 179–186. Study documenting how the Swedish Telecommunications Administration implemented automation within the telephone service company in the early 1960s.

"Halving the Workforce in a Petroleum Refinery," pp. 279–288. Study of the six-year (1957–1963) modernization effort of a large U.S. petroleum refinery. Modernization included construction of a new refinery, renovation of plants, and the installation of automatic and computer controls.

"Integrating Two Foundries," pp. 30–33. Study of an Austrian ironworks company during the merger of two obsolete factories into one remodeled, automated foundry in the early 1960s.

"Integrating Two Hose Units," pp. 234–241. Study of the modernization and expansion of a hose factory by a large British rubber company. Changes over the five-year period (1956–1961) included construction of a new hose factory with modern equipment and expansion of the product line.

"Introduction of EDP in a Canadian Insurance Company," pp. 36–45. Study of the effects of the installation of electronic data processing in a large Canadian insurance company from 1955 to 1961.

"Modernization and Shift Work in a Cotton Mill," pp. 227–233. Ten-year study (1954–1964) of the modernization of a millworks by a large British textile company. Major changes included the addition of modern machinery and a switch to continuous-shift production.

"Modernization and Staff Reduction in a Dyestuffs Plant," pp. 242–248. Study of the construction and staffing of a new, modernized British chemical plant (one of the largest in the industry) from 1957 to 1965.

"New Skills and the Older Worker — Problems of a Rolling Mill," pp. 151–157. Study of the replacement of an obsolete rolling mill in a Swedish ironworks (Avesta Jernvorks A.B.) with a modern foundry in the early 1960s.

"New Steelmaking Techniques and Computerized Control," pp. 215–226. Study of the effects of modernization of a large British steelworks from 1959 to 1965. Changes included installation of computers for production planning and control in a new steelmaking plant, a new primary rolling mill to replace the cogging mill, and a new strip-rolling mill.

"Personnel Planning a Major Need in Reorganization," pp. 102–110. Study of the reorganization and staffing of three plants of a large, West German metal-processing enterprise after the establishment of a new, modern finishing plant in 1964.

"Planned Rundown of Coal Industry's Labor Force," pp. 207–214. Study focusing on the human resource adjustments to modernization by the National Coal Board of the British coal industry from 1947 to 1963.

"Switch to Computer Accounting in Engineering," pp. 249–255. Study focusing on the job changes and staffing procedures following computerization of the accounting systems in a large British engineering firm. (No date given.)

"The Effects of External Forces on Staff Planning," pp. 111–120. Study of the reorganization and staffing of a large West German steel foundry subsequent to the building of two modern fine-plate rolling mills. Construction occurred during a period of labor surplus due to the German recession, but staffing took place during a period of severe labor shortages intensified by the building of the Berlin Wall.

"The Introduction of Four-Shift Working in Paper Manufacture," pp. 80–91. Three-year study (1959–1962) of the installation of new, more efficient machinery in a growing West German bulk paper mill.

"Time Lag in Human Adjustment to New Warehouse Methods," pp. 173–178. Study of the manpower adjustments involved in staffing a new, modern central warehouse at one of Sweden's largest merchandising firms in the early 1960s.

"Use of Redundant Employees in a New Paper-Container Factory," pp. 168–172. Study documenting the development of a modernized paper container factory by a large Swedish paper company (Iggesunds Bruk, Inc.) from 1960 to 1963. The technical equipment, supplied by American and English firms, was the first of its kind in Europe.

"Wholesale Switch to Electronic Data Bookkeeping in Sweden's Postal Bank," pp. 195–202. Study documenting the conversion from manual card punch to automated bookkeeping in the early 1960s within the central administrative office of a large Swedish banking concern (The Postal Bank in Stockholm).

Best, Ethel. "The Change from Manual to Dial Operation in the Telephone Industry." Washington, D.C.: U.S. Women's Bureau, 1933. Discussion of the impact of the switch from manual to dial operations in 1930 and the planning efforts of Bell Telephone Company to minimize worker dislocations.

Billings, R. S., R. J. Klimoski, and T. A. Breaugh. "The Impact of a Change in Technology on Job Characteristics: A Quasi-Experiment." *Administrative Science Quarterly* 22 (June 1977): 318–399. Discussion of the perceived and actual effects on job characteristics and satisfaction resulting from conversion to an assembly-line method of meal delivery in a large U.S. metropolitan hospital in the mid-1970s.

Bright, James R. *Automation and Management.* Boston: Harvard University Graduate School of Business Administration, 1958.

[*Automobile engine plant*]   Study of the effects on skill requirements and management responsibilities of four mechanized production processes in a large U.S. automobile engine plant (Ford's Cleveland engine plant) in the early 1950s.

[*Automobile manufacturer*]   Study of the organizational and employment effects resulting from the opening of a new plant for the automated plating of bumpers at a large U.S. automobile firm in the early 1950s.

[*Auto parts company*]   Discussion of the installation of an automatic work-feeding system in a manufacturing plant of a large U.S. engine parts firm in the early 1950s.

[*Bakery*]   Study of the impact on labor and management of the transition from three old bakeries to a new, highly automated bakery in the United States in the early 1950s.

[*Bumper-plating plant*]   Study of mechanization at a large automatic bumper-plating plant of a U.S. automobile company in the early 1950s.

[*Electrical machinery company*]   Study of the effects on employment of the introduction of seventy automatic assembly machines in the early 1950s in a large U.S. electrical machinery firm.

[*Engine plant*]   Study of the impact of the introduction of mechanized engine block assembly in a large engine plant of a U.S. automobile company in the early 1950s.

[*Feed mill*]   Discussion of the design, installation, operation, and labor requirements of a new, small, highly automated feed mill in the United States in the mid-1950s.

[*Fertilizer company*]   Study of the organizational and personnel changes resulting from the mechanization of a rebuilt plant in a small U.S. fertilizer company.

[*Metalworking job shop*]   Discussion of the impact of automation of the production line for oil seals for various kinds of machinery at a U.S. metalworking firm in the 1950s.

[*Mining company*]   Discussion of the experience of a small U.S. mining company's decision to automate in the early 1950s and return to the more profitable deep mining instead of strip mining.

[*Oil refinery*]   Discussion of the organizational and labor requirements of a small, new U.S. integrated oil refinery, which used an unusually high degree of centralized automatic control, in 1952.

[*Rubber products company*]   Study of the organizational and employment changes after mechanization of handling, work flow, and storage in the production of mattresses and cushions in a large U.S. rubber products factory in the mid-1950s.

Buchanan, David A., and David Boddy, "Advanced Technology and the Quality of Working Life: The Effects of Word-Processing on Video Typists." *Journal of Occupational Psychology* 55, no. 1 (March 1982): 1–11. Description of the impact of word-processing technology on typing jobs in a Scottish engineering consultancy in the late 1970s.

Canadian Department of Labor. *Technological Changes and Skilled Manpower: Electronic Data Processing Occupations in a Large Insurance Company*. Ottawa: Research Program in the Training of Skilled Manpower, 1961. Detailed description of the planning, occupational changes, staffing practices, and training associated with a large Canadian insurance company's conversion to electronic data processing in the 1950s.

"Computer und Angestellte," Industriegerverkschaft Metall für die Bundesrepublick Deutschland, Oberhausen (March 5–8, 1968).

[*Electrical engine firm*]   Summary of the effects of the introduction of electronic data processing in the order department in a medium-sized West German electrical engineering firm in the 1960s.

[*Electrical engineering firm*]   Summary of the effects of the introduction of electronic data processing in a research department of a large West German electrical engineering firm in the 1960s.

[*Metalworking plant*]   Summary of the effects of the introduction of electronic data processing in the payroll office of a large West German nonferrous metal-producing firm in the 1960s.

[*Sheet steel producer*]   Summary of the effects of the introduction of electronic data processing for production and stock planning in a sheet steel mill in West Germany in the 1960s.

[*Smelting works, production planning*]   Summary of the effects of the introduction of electronic data processing for production planning in a large West German smelting works in the 1960s.

[*Smelting works, stock department*]   Summary of the effects of the introduction of electronic data processing in the stock department of a large West German smelting works in the 1960s.

[*Special engineering firm*]   Summary of the effects of the introduction of electronic data processing for stock-keeping and purchasing functions in a medium-sized, special engineering firm in West Germany in the 1960s.

[*Steel and plant construction firm*]   Summary of the effects of the introduction of electronic data processing for financial, cost, and property-accounting functions in a West German steel and plant construction firm in the 1960s.

[*Transportation equipment company, payroll office*]   Summary of the effects of the introduction of electronic data processing in the payroll office of a large West German transportation equipment company in the 1960s.

[*Transportation equipment company, testing division*]   Summary of the organizational and employment effects of the introduction of electronic data processing in the works laboratory and materials-testing division of a large West German transportation equipment manufacturer in the 1960s.

Connor, R. W., and R. W. J. Rothwell. "E.D.P. in a Public Utility—Effects on Organization, Employment and the Personnel Function." *Three Computer Case Studies*. Melbourne, Australia: Department of Labour and National Service, 1971. Discussion of the organizational and personnel changes resulting from the introduction of electronic data processing over a seven-year period (1959–1966) in a large, privately owned public utility in Australia.

"Coping with Technological Change: Carrington Vizella's Experience." *Industrial Relations Review and Report* 222 (April 1980): 2–5. Brief discussion of the effects on employment, training, and working practices of the replacement of two conventional British textile mills with one modern plant in the late 1970s. The new mill is equipped with electronic controls and operated in a continuous shift system. [Supplemented by case study of Carrington Vizella Yarns Ltd. in Income Data Services Ltd., "Introducing New Technology," IDS Study 202 (London, June 1980.]

Cossey, Hubert, Lizette Cazmar, and George Hedebouw. "Innovation in Shiftwork Systems in the Textile Sector." *Shiftwork in the Textile Industry—Case Studies of Innovations*. Belgium: European Foundation for the Improvement of Living and Working Conditions, April 1981. Discussion of the opening of a technologically advanced spinning department in a Belgian textile firm in the 1970s. The department was created in 1972 after a "crisis" in the industry attributed to keen competition from "low-pay" countries.

Craig, Harold. "Administering Technological Change in a Large Insurance Office—A Case Study." *Industrial Relations Research Association Proceedings* (1954), pp. 129–138. Study of the effects of automation on the staff in the head office of a large U.S. insurance company in the early 1950s.

Crossman, Edward R. F. W., and Stephen Laner. *The Impact of Technical Change on Manpower Skill Demand: Case-Study Data and Policy Implications*. Berkeley: University of California, 1969.

[*Air products firm*] Study of the effects of the installation of automatic, computer-monitored control devices in a small U.S. air separation plant in the mid-1960s.

[*Bank*] Detailed study of the effect on skill requirements, employment, and unemployment of the introduction of electronic data processing in the branch offices of a large U.S. bank (Bank of America) in the late 1960s.

[*Domestic airline*] Study of the conversion from manual to semi-automatic data-processing operations in the airline reservations system of American Airlines in the mid-1960s.

[*Electric power plants*] Study of the impact of the installation of computerized process control on two U.S. electric power plants in the mid-1960s.

[*Electric power plants*] Study of the effects of the introduction of new turbine generator boiler units and centralized control rooms in three electric power plants in the U.S. in the mid-1960s.

[*International airlines*] Study of the effects of the conversion from semi-automatic to electronic data processing in the reservations system of Pan American Airlines in the mid 1960s.

[*Petroleum refinery*] Study of the effects of the introduction of computer-ization on the automatic process control functions in two plants of a U.S. petroleum refinery in the mid-1960s.

[*Steel-processing plant*] Study of the effects of conversion from an older batch-annealing process to a newer, continuous process in a U.S. steel facility in the mid-1960s.

Cruz, Daisy. "Affirmative Action at Work." *Personnel Journal* 55, no. 5 (May 1976): 226ff. Description of the staffing and training for a new department responsible for a composition operation in the production of tactical missiles in a large U.S. aerospace firm (Lockheed) in the mid-1970s.

Dickson, Keith. "Petfoods by Computer: A Case Study of Automation." In Tom Forester, ed., *The Microelectronic Revolution*. Cambridge, Mass.: MIT Press, 1980, pp. 174–183. Detailed account of the impact of the introduction of automated food processing and canning facilities in production, organiza-tion, skill requirements, and management in a British pet food factory in 1978.

Diebold Institute for Public Service Studies. *Labor Management Contracts and Technological Change*. New York: Praeger, 1969.

"Armour and Company and the Amalgamated Meat Cutters and Butcher Workmen and the United Packinghouse Workers of America," pp. 38–55. Study of the ways in which management and labor in a large meat-packing company attempted to solve the problem of labor displacement brought about by modernization of the company's meat-processing and slaughtering plants, and by the diversification of its production activities. Focuses on the 1960s.

"Kaiser Steel Corporation and the U.S. Steelworkers of America," pp. 25–37. Study of the effects of technological change on employment and collective bargaining during the 1960s at an integrated steel mill owned by the nation's ninth-largest U.S. steel producer.

"Pacific Maritime Association and the International Longshoremen's and Warehousemen's Union," pp. 56–71. Study of labor-management rela-tions during the modernization and mechanization of longshoring pro-cesses in the shipping industry on the west coast, primarily in the 1960s. Major changes involved installation of more efficient, more sophisticated machinery for handling and containerization.

Eastman, Susan. "E.D.P. in a Manufacturing Company." *Three Computer Case Studies*. Melbourne, Australia: Department of Labor and National Service, 1971. Detailed account of a large Australian manufacturing company's progression from punch-card facilities to a small computer to a larger computer in the 1960s.

Eckert, Susan. "Continuous Technological Advances Require Continuing Em-ployee Education." *Telephony* (November 9, 1981): 26ff. Study of the em-ployment and educational needs resulting from the installation and use of computers at Rochester Telephone Company.

Erivansky, Arkadii. *A Soviet Automatic Plant*. Moscow: Foreign Language Publishing House, 1955. Study of the application of automated technology in a Soviet piston production plant of an automobile company in the mid-1950s. Automation extended to various production processes, remote controlling, casting, product testing, and safety enhancement.

Faunce, William A. "Automation and the Automobile Worker." *Social Problems* 6, no. 1 (Summer 1958): 68–78. Discussion of the nature of job changes and their effects on workers resulting from automating the assembly line in a large engine plant of a U.S. automobile company in the mid-1950s. [Supplemented by William A. Faunce, "The Automobile Industry: A Case Study in Automation," in Howard B. Jacobson and Joseph S. Roucek, eds., *Automation and Society* (New York: Philosophical Library, 1959), pp. 44–53.]

Greenhalgh, N. H., "Automation at Gawith's Bakery." *Personnel Practice Bulletin* 18, no. 2 (June 1962): 27–32. Discussion of staffing, training, and organizational changes relating to the introduction of automated equipment in an Australian bakery employing fifty-six.

Gunzburg, D. "A Computer in an Office — Organization and Personnel Effects." *Three Computer Case Studies*. Melbourne, Australia: Department of Labor and National Service, 1971. Detailed account of the impact of electronic data processing on the organizational structure, occupational changes, and personnel policies at a large Australian insurance company (National Mutual Life Association).

Haddy, Pamela. "Some Thoughts on Automation in a British Office." *Journal of Industrial Economics* 6 (1958): 161–170. Discussion of the development and introduction of a new, high-speed electronic computer in a large British restaurant services firm (J. Lyons & Company) in the late 1940s.

Hardin, Einer. "Computer Automation, Work Environment and Employee Satisfaction: A Case Study." *Industrial and Labor Relations Review* 13, no. 4 (1959): 559–567. Study of the effects of installation of electronic data processing in the home office of a medium-sized casualty insurance company in the United States in the mid 1950s.

Incomes Data Services, Ltd. "Introducing New Technology." IDS Study 202. London, June 1980.

[*Bradford & Bingley Building Society*]   Brief summary of the effects on employment and output of the installation in 1977 of word-processing equipment in the offices of one of the largest building societies in Great Britain.

[*British Leyland — Mini Metro Plant — Longbridge*]   Brief summary of the effects on employment and output following introduction of fully automated, robot assembly line in a British automobile plant in the late 1970s.

[*British Sugar Corporation*]   Brief summary of the impact of the introduction of microprocessor controls into the sugar-production process in one of seventeen plants owned by a large sugar beet processor in the late 1970s.

[*City of Bradford Metropolitan Council — Yorkshire*]   Brief summary of the impact of the introduction of word-processing equipment in a local government administrative office in the late 1970s.

[*ECM Engineering Designs*]   Brief summary of the effects of the introduction of microprocessor control systems into a small British machinery company's line of guillotines (papercutters) in the late 1970s.

[*Halle Models Ltd.*] Brief summary of the impact of the introduction of electronic equipment into the manufacture of ladies' and children's clothing in a 360-person factory in the United Kingdom in the late 1970s.

[*London Penta Hotel*] Brief discussion of the effects on workers and jobs of the introduction of an integrated computer system to assist in reception, reservation, and accounting tasks in a large British hotel.

[*Thorn-EMI Consumer Electrics Ltd.*] Brief discussion of the impact on skill requirements and employment of the introduction of computerized control and assembly equipment in a large television-manufacturing plant of a British consumer electronics firm.

Incomes Data Services, Ltd. "Introducing New Technology." IDS Study 252. London, 1981.

[*Cadbury Schweppes Ltd.*] Summary of the employment effects of installation of word-processing equipment in a large office in the United Kingdom in 1978.

[*Colonial Mutual Life Assurance Society Ltd.*] Study documenting the results of reorganization of a typing center and the introduction of word processors in the head office of a large British insurance company from 1978 to 1981.

[*Crusader Insurance Company, Ltd.*] Study documenting the installation and staffing of word processors and minicomputers in a medium-sized British insurance company in the late 1970s.

[*London Transport Executive*] Study documenting the installation of word processors and computer terminals from 1974 to 1981 in a large London transportation services management group.

[*Norwich Union Insurance Group*] Study documenting the gradual introduction of word processors and minicomputers at a large British insurance company from 1972 to 1980.

[*Richard Costain Ltd.*] Study documenting the introduction and staffing of word processors at the head office of a large British construction company in 1979 and 1980.

"Industrial Robots: Social Effects in Automobile Manufacturing." *Social and Labour Bulletin.* (December 1982): 444–446. Brief description of the effects of the introduction of robots on the quantity and quality of jobs at a large West German automobile manufacturer.

Jacobsen, Eugene; Don Trumbo; Gloria Check; and John Nagle. "Employee Attitudes Toward Technological Change in a Medium-Sized Insurance Company." *Journal of Applied Psychology* 43, no. 6 (December 1959): 349–354. Results of a survey of office workers regarding the changes in their jobs after installation of an electronic computer in a medium-sized U.S. insurance company in 1956.

Jacobson, Howard Boone, and Joseph S. Roucek, eds. *Automation and Society.* New York: Philosophical Library, 1959.

Bailey, S. B., discussion of Crosley-Bendix (Division of AVCO), appendix. Brief look at the effects on employment of new, automated equipment introduced in a steel-cutting process in a large U.S. household appliance firm in the mid-1950s.

Bailey, S. B., discussion of Electrolux Corp., appendix. Brief look at the impact of the introduction of an automated coil-winding process in a large U.S. appliance parts firm in the mid-1950s.

Bailey, S. B., discussion of Ford Motor Company, appendix. Study documenting the effects of major automation changes in an assembly plant and a stamping plant of a large U.S. automobile firm in the mid-1950s. Changes included the opening of a modern plant and the introduction of automated spot-welding systems and new loading equipment.

Bailey, S. B., discussion of Macmillan & Bloedel, appendix. Brief discussion of the effects on employment of automation in the plywood-manufacturing division of a large Canadian lumber company in the mid-1950s. Changes mainly involved automatic rolling, pressing, and gluing operations.

Beirne, Joseph A., "The Nature of Automation in the Telephone Industry," pp. 183–192. Discussion of the employment effects (as viewed by the Communications Workers of America) of various technological changes in AT&T from 1920 to 1954. Major changes included the introduction of automatic message accounting, direct dialing, and new automatic electronic test equipment.

Cordiner, Ralph J., "Automation in the Manufacturing Industries," pp. 19–33. General overview of the effects of improving methods, mechanizing, and automating the production processes of a large U.S. electrical equipment manufacturer (General Electric) over twenty years (1939–1958).

Davis, D. J., "Automation in the Automotive Industry," pp. 34–43. Study documenting the introduction of automated processes at an engine plant of a large U.S. automobile manufacturing firm (Ford Motor Company) from 1950 to 1959. Major changes included mechanized handling equipment, automated control panels, and automated press lines.

Jansky, Curtis, "Automation in Data Processing for the Small-to-Medium-Sized Business," pp. 145–153. Study of the conversion to automated data processing of the sales analysis department in a medium-sized U.S. manufacturing firm in the late 1950s.

Mitchell, Dan G., "Automation in the Electronics Industry," pp. 68–81. Study highlighting the importance of automation and technological changes for economic and employment growth in a large U.S. electronics firm (Sylvania Electric Company) from the 1930s through the 1950s.

Osborn, David G., "Automatic Data Processing in the Large Company," pp. 154–172. Study of the two-year conversion process to automated data processing in a large U.S. manufacturing firm in the late 1950s.

Phalen, C. W., "Automation and the Bell System," pp. 173–182. Study of the employment effects (as viewed by AT&T) of various technological changes in the telephone company from 1920 to 1954. Major changes included automatic message accounting and conversion to automated direct-dial service.

Jakubauskas, Edward B., "Adjustment to an Automatic Airline Reservation System." Report 137. Washington, D.C.: U.S. Department of Labor, Bureau of Labor Statistics, 1958. (Reprinted in BLS Bulletin 1287, November 1960.) Description of job changes and training requirements due to the adoption of an electronic reservations system at a major U.S. airline in 1952.

Karbowski, Yvonne S. "A Meat Packing Plant Mechanizes — Omaha's Response to Automation Layoffs." *Employment Security Review* 29, no. 7 (July 1962): 34–36. Discussion of the disposition of workers laid off when alarge meat-packing company (Cudahy Meat-Packing Company of Omaha) replaced an old plant with a new, automated facility in the early 1960s.

Lester, Tom. "Tale of a Toolmaker." *Management Today* (July 1981): 52–57. Brief description of the effects of installation of state-of-the-art toolmaking equipment in a small, British tool-manufacturing firm in the 1970s.

Lipstrev, Otis, and Kenneth A. Reed. *Transportation to Automation: A Study of People, Production and Change.* Boulder: University of Colorado Press, 1964. Study of the effects of change from old mill-type factories to very modern production facilities at a large bakery owned by a large midwestern baking company in the early 1960s.

Mann, Floyd C., and Richard L. Hoffman, *Automation and the Worker: A Study of Social Change in Power Plants.* New York: Holt, 1960. Study of the effects of a newly built, technologically superior electric power plant on an old, less efficient plant during the 1950s.

Mann, Floyd C. and Franklin W. Neff, *Managing Major Change in Organizations.* Ann Arbor, Mich.: Foundation for Research and Human Behavior, 1961.

[*Health insurance company*]   Study of the installation of high-speed electronic data-processing equipment in a U.S. health insurance company in the 1950s.

[*Steel pipe-manufacturing company*]   Study of the employment and organizational effects of a U.S. steel pipe manufacturer's decision to build a highly modernized seamless pipe mill in the early 1950s.

Mann, Floyd C., and Laurence K. Williams. "Observations on the Dynamics of a Change to Electronic Data Processing Equipment." *Administrative Science Quarterly* 5, no. 2 (September 1960): 217–256. Study of the effects of a change to electronic data-processing equipment in a large U.S. electric power and light company. Change was introduced over a five-year period and primarily affected the accounting and sales divisions.

Maryland Department of Employment Security. *The Impact of Technological Change in the Banking Industry.* U.S. Employment Service, Automation Manpower Services Program, June 1967. Description of the personnel changes following the installation of electronic data-processing equipment in a large commercial bank (in Baltimore) in 1961.

Menzies, Heather. *Women and the Chip: Case Studies of the Effects of Informatics on Employment in Canada.* Montreal: Institute for Research on Public Policy, 1981.

[*Life insurance company*]   Study of the impact of information automation or informatics on a large Canadian life insurance company in the late 1970s and early 1980s.

[*Transportation and communications company*]   Study depicting the impact of the application of microelectronics in the head office of a large Canadian corporation in the transportation and communications sector in the late 1970s.

Miller, Ben. "Gaining Acceptance for Major Methods Changes." Research Study No. 44. New York: American Management Association, 1960.

[*Commercial bank*]   Study of the conversion from manual to automated check processing in the offices of an expanding commercial bank in the late 1950s.

[*Commercial bank*]   Study of the conversion from a manual to an automated bookkeeping system in the loan department of the main office of a large U.S. commercial bank in the late 1950s.

[*Paper products firm*]   Study of the conversion of manual recordkeeping to a card-punch method in the late 1950s in the administrative office of a large American paper products firm.

[*Pharmaceutical firm*]   Study of the conversion from manual to automated sales analysis in the administrative office of a large U.S. pharmaceutical firm in the late 1950s.

[*Public utility*]   Study of the conversion from manual to automated customer billing services in a large U.S. public utility company in the late 1950s.

[*Telephone company*]   Study of the conversion from manual to electronic billing procedures in the late 1950s in the head office of an independent American telephone company.

Mumford, Enid, and Olive Banks. *The Computer and the Clerk*. London: Routledge and Kegan Paul, 1967.

[*Bank*]   Discussion of the effects of the installation of a computer in the office of a British bank in the late 1950s.

[*Cattle food company*]   Discussion of the effects of the installation of a computer in the office of a medium-sized British cattle food company in the late 1950s.

"No Shuttlecocks at Parlin." *Fortune* (February 1961): 189–190. Brief discussion of the employment effects resulting from implementation of new, automated machinery and continuous shift production at a U.S. photographic film plant (Dupont) in 1958.

Online Conference Limited. *Word Processing, Selection, Implementation and Uses*. Uxbridge, England, 1979.

Cahill, Tim, "Staffordshire County Council's Text Processing System," pp. 235–242. Description of the installation of a text-processing system in a local British government office in the mid-1970s (County Clerk and Chief Executives, Department of Staffordshire County Council.)

Day, Janet, "How a Solicitor Uses Word Processing—A Case Study," pp. 243– 252. Description of the opening of a word-processing department at a British legal services firm in the mid-1970s.

Edson, Mariane B., "Managing the People and the System," pp. 107–114. Discussion of the impact of the introduction of word-processing equipment on the typing staff and overall organization of a scientific research division of a British agrochemical firm in the mid-1970s.

Parnham, Alan, "Word Processing at Willis, Faber and Dumas," pp. 229–234. Brief description of the introduction of word-processing equipment in a large British insurance company in the late 1970s.

Spooner, Robert, "Word Processing for an Expanding Agent's Service," pp. 253–260. Description of the selection process and impact of installation of word-processing equipment in a small British real estate firm in the mid-1970s.

Organization of Economic Cooperation and Development (OECD). *Adjustment of Workers to Technical Change at the Plant Level: Supplement to the Final Report.* Paris, 1966.

Cassell, Frank H., "Corporate Manpower Planning and Technical Change at the Plant Level," pp. 256–267. Report on the effects of the impact of automation at a large U.S. steel company (Inland Steel Company) in the 1950s and 1960s. Major changes include introduction of four continuous galvanizing lines and a basic oxygen furnace and the opening of a computerized eighty-inch hot strip mill.

Louet, Roger, "Planning Coordination: The Saint Nazaire Shipyards," pp. 207–218. Report on the social and economic consequences of a large-scale reduction in the work force employed at a large French shipyard (Atlantic Shipyards) from 1958 to 1966. The shipyard was modernized and reorganized to save the ailing company during a period of secular decline in the French shipbuilding industry.

Mitchell, Sir Stuart, "Planning Coordination: British Railways," pp. 184–196. Description British Railways' massive reduction of its work force after technological changes in its main workshops in the early 1960s.

Sasaki, Hideichi, "Planning Coordination: The Sasaki Glass Company, Ltd.," pp. 203–206. Brief look at the adjustment of job and wage structures to technological change in a Japanese glass company from 1956 to 1966.

Willener, A., "A French Steelworks – Closure of an Old Rolling Mill: Problems of Co-ordinated Transfers of Personnel," pp. 145–158. Study describing the methods and problems associated with the transition to a modernized French steelworks that replaced an older, smaller steelmaking shop in the mid-1950s.

Willener, A., "Manpower Planning and Redeployment in an Expanding Business," pp. 159–173. Description of the resettlement and coordination of a new, automated production department in a large French manufacturing firm.

Organization of Economic Cooperation and Development (OECD). *Office Automation: Administrative and Human Problems.* Paris, 1965.

Eliaeson, P. J., "A Smooth Transition and Reduced Labor Force in Swedish Insurance Companies," pp. 55–67. Study of the effects of the introduction of electronic data processing on the work force and jobs in two insurance companies in Sweden around 1960.

Marenco, Claudine, "Gradualism, Apathy and Suspicion in a French Bank," pp. 31–53. Discussion of changes in employment, jobs, recruitment, and training following computerization of risk analyses and bill processing in the administrative centers of a large, nationalized French bank in the late 1950s and early 1960s.

Osborn, Roddy F. "GE and UNIVAC: Harnessing the High Speed Computer." *Harvard Business Review* 32, (July/August 1954): 99–107. Description of the installation and application of a high-speed computer at a large U.S. electrical equipment manufacturer in 1953 and 1954.

Paul, B. R. "The Introduction of Electronic Data Processing in Life Assurance." *Personnel Practice Bulletin* 18, no. 2 (June 1962): 7–11. Discussion of the effects of the introduction of electronic data processing on employment, recruitment, and training in a large life insurance firm (the A.M.P.) in Australia in the early 1960s.

Pennsylvania State Employment Service. *The Effect of Automation on Occupations and Workers in Pennsylvania.* May 1965.

[*Bituminous coal mine*] Brief description of the effect on labor requirements of the introduction of a completely mechanized production process (the Wilcox continuous-mining method), at a bituminous coal mine in the United States in the late 1950s.

[*Fabricated metal products firm*] Brief description of the impact of new, automatic machinery in the finishing operations in a U.S. firm manufacturing collapsible metal tubes in the early 1960s.

[*Furniture company*] Brief description of the change in skill requirements resulting from installation of an automated conveyor line in the finishing department of a small U.S. manufacturer of high-quality wooden desks in the early 1960s.

[*Limestone quarry*] Brief description of the effect on labor requirements of the introduction of a completely automated production operation in a new U.S. stone-crushing quarry in 1963.

[*Mechanical power transportation equipment company*] Brief description of the effects on the work force of the introduction of a new automatic press at a mechanical power transmission equipment firm in the United States.

[*Mining machine and equipment company*] Brief description of the effects on the work force of installation of numerically controlled tape machines in a U.S. mining machine and equipment firm.

[*Public utility*] Brief description of the impact of the introduction of an automated data-processing center and a supervisory control system in the pumping stations of a city water department in the United States in 1959.

[*Rubber company*] Brief description of changes in skill requirements and employment upon introduction of new curing machines in a rubber tire- and tube-manufacturing firm in the United States.

Riche, Richard W., and William E. Alli. "Office Automation in the Federal Government." *Monthly Labor Review* (September 1960): 109–113. Review of congressional hearings in 1959 and 1960 on the extent of office automation in the federal government and the effects of technological changes on clerical employees. Focuses on the installation of computers in the Department of the Treasury and Veteran's Administration.

Riche, Richard W., and James R. Alliston. "Impact of Office Automation in the IRS." *Monthly Labor Review* 86, no. 4 (April 1963): 388–393. Study highlighting the human resources implications of the conversion in the early 1960s to a comprehensive, automatic data-processing system within the Internal Revenue Service and its initial introduction in the Atlantic Regional Division.

Rothberg, Herman J. "Adjustment to Automation in a Large Bakery." Bulletin 1287. Washington, D.C.: U.S. Department of Labor, Bureau of Labor Statistics, 1956. Discussion of the human resource adjustments made with the opening of a new, centralized mechanized bakery in the United States in 1953.

Rothberg, Herman J. "Labor Adjustments for Changes in Technology at an Oil Refinery." In Bulletin 1287. Washington, D.C.: U.S. Department of Labor, Bureau of Labor Statistics, 1960. Study of the effects of technological changes on the workplace and work force from 1948 to 1956 in a medium-sized U.S. oil refinery. Major changes included installation of a fluid catalytic cracking unit, a delayed coking unit, a new crude distillation unit, and a new catalytic reforming unit.

Rothwell, Roy, and Walter Zegveld. "National Coal Board." In *Technical Change and Employment*. New York: St. Martin's Press, 1979, pp. 66–76. Discussion of employment effects of postwar mechanization on the National Coal Board in the United Kingdom.

Salmans, Sandra. "Pilkington's Progressive Shift." *Management Today* 163 (September 1980): 66–73. Study of the effects of the installation of a highly automated, manufacturing process (the float process) in a large, multinational glass-manufacturing company based in Great Britain in the late 1970s.

Schultz, G. P., and T. L. Whisler, eds. *Management, Organization and the Computer*. Glencoe, Ill.: Free Press, 1960.

[*Atwood Vacuum Machine Company*]   Discussion of the impact of replacing a punch-card batch system with a computer specifically geared toward random-access accounting and control in the main office of a large U.S. auto parts firm in the late 1950s.

[*International Shoe Company*]   Discussion of the effects of management and jobs on the introduction of an electronic computer in a large U.S. shoe manufacturer in the late 1950s. Computers were used primarily for inventory control and production requirements and scheduling.

[*Standard Oil of New Jersey*]   Discussion of the impact of the installation of computers and electronic data-processing equipment for use in production, transportation, manufacturing, marketing, engineering, and research functions in a large U.S. oil company in the 1940s and 1950s.

Scott, W. H.; J. A. Banks; A. H. Halsey; and T. Lupton. *Technological Change and Industrial Relations*. Liverpool, England: Liverpool University Press, 1956. Study of the effects of a switch to continuous production and improvements in stripping and melting procedures in a large steel mill located in northern Wales in the mid-1950s.

Sheppard, Harold L., and James L. Stern. "Impact of Automation on Workers in Supplier Plants." *Labor Law Journal* 8 (1957): 714–718. Brief discussion of the automation of the stamping process in a large American automotive firm in 1947 and the subsequent closing of a supplier plant. Major changes included the installation of new presses equipped with automatic loading and unloading devices, automatic transfer equipment, and new high-speed, automatic welding machines.

Stanford Research Institute. *Management Decisions to Automate.* Adapted by Harry F. Bonfils. Washington, D.C.: Office of Manpower, Automation and Training, U.S. Department of Labor, 1965.

"Automation at the Hamilton Electronics Company," pp. 22–24. Brief discussion of the effects on employment of the installation of numerically controlled wiring equipment in a large U.S. electronic equipment-manufacturing firm in the early 1960s.

"Automation at the Marlin Distributing Company," pp. 24–25. Brief summary of the effects on employment of automated order-assembly machinery in the warehouse of a U.S. consumer products distributing company in the early 1960s.

[*Bank*]   Study of the effects of a large, rapidly growing bank's conversion to automated electronic data processing in 1963 and 1964.

[*Branch bank*]   Study of the effects of a growing branch bank's conversion to an electronic data-processing system in the United States from 1957 to 1962.

[*Consumer products firm*]   Brief overview of the effects on employment of installation of an automatic order assembly system in a warehouse of a U.S. consumer goods manufacturer and distributor in the late 1950s and early 1960s.

"The Milwaukee-matic at Alger Electronics," pp. 21–22. Brief summary of the effects of the introduction of numerically controlled, multipurpose machinery in the production of electronic testing equipment in a large U.S. manufacturing firm.

"Warehouse Automation at the V.C.P. Company," pp. 25–26. Brief discussion of the changes in employment resulting from the installation of automatic assembly systems in a frozen-food warehouse in the United States in the early 1960s.

Thakur, C. P., and G. S. Aurora. *Technological Change and Industry.* New Delhi Shri Ram Centre for Industrial Relations, 1971.

Hein C. Jain. "Technological Changes in the Oil Industry," pp. 330–336. Detailed discussion of the impact of computerization on employment in an oil company in India in the 1960s.

E. P. Mathias, "Management of Technological Change in Railways," pp. 101–133. Detailed account of the effects on employment and staff of modernization of the Indian railway system. Changes included the introduction of route relay interlocking stations, diesel engines, and computerization of accounting and control tasks.

T. S. Papola, "On Introducing Technological Change: A Study of Ahmedabad Cotton Textile Industry," pp. 245–264. Discussion of the problems associated with the replacement of an old loom shed with a new one (Gapal Cotton Mills Ltd.) equipment with automatic looms in India in the early 1950s.

M. V. Pyles, and P. R. Podwal. "Technology, Organization, and Industrial Relations: A Case Study," pp. 87–100. Discussion of the introduction of a new technology in the production of ammonia salt (1947) and in a new oil gasification plant (1962) in a chemical company (Fertilizer and Chemicals Travancore Ltd.) in India.

C. P. Thakur, "Labour Market and New Technology: A Study in Ranchi." pp. 209–228. Discussion of skill requirements and recruiting for a large, modern factory of a heavy machine and equipment manufacturer in India in the 1960s.

U.S. Civil Service Commission. *The Impact of Automation on Federal Employees.* Washington, D.C.: Government Printing Office, 1974. Reports resulting from a study on the impact of automation on employment and the work force in twenty U.S. government agencies, employing over 2 million workers, from 1961 to 1963.

U.S. Department of Labor, Bureau of Labor Statistics. "A Case Study of a Company Manufacturing Electronic Equipment," Studies of Technology, No. 1. Washington, D.C., October 1955. Study of the effects of automated production techniques in the manufacture of electronic equipment in one of the largest producers of radio and television sets in the United States in the mid-1950s.

U.S. Department of Labor, Bureau of Labor Statistics. *Impact of Technological Change and Automation in the Pulp and Paper Industry,* Bulletin No. 1347. Washington, D.C., October 1962.

"Case Study of an Automatic Paper Roll Handling System," pp. 67–77. Study of the effects of the installation of an advanced, automatic roll-handling system in a modernized plant of a large U.S. paper manufacturer in the 1950s.

"Case Study of the Introduction of Continuous Processing Equipment," pp. 50–66. Study of the effects of the installation of a continuous digester to replace old equipment in a large pulp and paper plant in the mid-1950s.

"Case Study of Mechanization of Materials Handling," pp. 31–49. Study of the effects of mechanization of handling operations in a large U.S. pulp and paper mill in the mid-1950s.

U.S. Department of Labor, Bureau of Labor Statistics. *Industrial Retraining Programs for Technological Change — A Study of the Performance of Older Workers.* Bulletin 1368. Washington, D.C., June 1963.

"Retraining of Aircraft Production Workers, Technicians and Engineers," pp. 13–20. Study focusing on the retraining efforts resulting from changes in aircraft production due to rapid advances in weapons systems and electronic technology in a large U.S. aircraft manufacturer in the early 1960s.

"Retraining of Oil Refinery Production Workers," pp. 7–12. Discussion of the retraining of workers brought about by installation of highly integrated, continuous production and instrumentation control processes in a relatively small U.S. oil refinery in the early 1960s.

"Retraining of Telephone Operators," pp. 25–31. Discussion of the retraining of several hundred telephone operators due to the installation of electronic data-processing systems in the accounting department of a U.S. telephone company in the early 1960s.

U.S. Department of Labor, Bureau of Labor Statistics. "The Introduction of an Electronic Computer in a Large Insurance Company." Studies of Automatic Technology, Case 2. Washington, D.C.:, October 1955. Study of job changes and staffing needs resulting from the introduction of a computer in the classification department of a large U.S. life insurance company in 1955.

van Auken, K. G., Jr. "Plant Level Adjustments to Technological Change." Monthly Labor Review 76, no. 4 (April 1953): 388–391.

[Ladies' blouse manufacturer] Summary of study of the effects of installation of new buttonholing machines in the early 1950s in a small U.S. blouse manufacturer.

[Ladies' garment manufacturer] Summary of study on how workers and management in a ladies' garment-manufacturing plant in the United States adjusted to the introduction of machine trimmers in the early 1950s.

[Ladies' slip manufacturer] Summary of study on how workers and management in a small U.S. manufacturer of women's slips adjusted to the introduction of four automatic presses in 1951.

Walker, Charles R. Toward the Automatic Factory: A Case Study of Men and Machines. New Haven, Conn.: Yale University Press, 1957. Discussion of the employment requirements of the first continuous seamless pipe mill in a large U.S. steel company (U.S. Steel) from 1949 to 1952.

Waters, Craig R. "There's a Robot in Your Future." Inc. 4, no. 6 (June 1982): 64–74. Discussion of the introduction of robots in a small U.S. die-casting plant (Newton-New Haven Co.) around 1980.

Weber, C. Edward. "Change in Managerial Manpower with Mechanization of Data Processing." Journal of Business (April 1959): 151–163.

[Manufacturing firm] Study of the effects of the installation of electronic data processing on office personnel in a large U.S. manufacturing firm in the late 1950s.

[Steel company] Study of the effects of the installation of electronic data processing on office personnel in a basic steel company in the United States in the late 1950s.

Weingard, Marvin. "The Rockford Files: A Case for the Computer." Management Accounting (April 1979): 36–38. Study of the conversion to electronic data processing of a medium-sized paperboard- and carton-manufacturing firm (Rockford Paper Mills, Inc.) in the late 1970s.

Wilkins, Russell. Microelectronics and Employment in Public Administration: Three Ontario Municipalities, 1976–1980. Ontario: Ministry of Labor, July 1981.

"Informatics Technology and Employment in a Large City: Toronto," pp. 19–46. Study of the impact of computerization in the municipal offices of a large city in Canada from 1974 to 1979.

"Informatics Technology and Employment in a Medium-Sized City: Ottawa," pp. 13–18. Study of the impact of computerization in the municipal offices of a medium-sized city in Canada from 1967 to 1980.

"Informatics Technology and Employment in a Small Metropolitan Centre: Oshawa," pp. 7–12. Study of the impact of computerization in the municipal offices of a small city in Canada from 1970 to 1980.

Wilkinson, Barry. "Managing with New Technology," *Management Today* (October 1982): 33–40. Discussion of the effects on work skills and organization of computerization of the blank selection, tool selection, and machine settings in the manufacture of lenses and spectacles at a small optical firm (Derby Optical Co.) in the United Kingdom.

Wisconsin State Employment Services. "A Large Insurance Company Automates: Workforce Implications of Computer Conversion." Automation Manpower Services Program Demonstration Project No. 3. Madison, April 1964. Detailed study of the work force implications of computer conversion in a large American insurance company in the early 1960s.

Yavitz, Boris. *Automation in Commercial Banking*. New York: Columbia University Graduate School of Business and The Free Press, 1967.

[*Essex County National Bank*] Study of the effects of computerization on jobs and working conditions in a medium-sized U.S. commercial bank in the mid-1960s.

[*Fidelity Bank and Trust Company*] Study of the effects of computerization on jobs and working conditions in a large U.S. commercial bank in the mid-1960s.

[*Manufacturers National Bank*] Study of the effects of computerization on jobs and working conditions in a small U.S. commercial bank in the mid-1960s.

[*Metropolitan Bank and Trust Company*] Study of the effects of computerization on jobs and working conditions in a large U.S. commercial bank in the mid-1960s.

Index

advisory committees, 115
agglomeration economies: and
  emerging industries, 183n.1; and
  high technology industries, 5; and
  the international product life
  cycle, 166n.8; and Route 128, 89
aggregation problems, in
  determining impacts of
  technological change, 6, 45
allocation of work, 21, 38–44,
  148–149. *See also* employer
  adjustments
apparel industry, 84–85; 95–96
apprenticeship programs, 33, 69
associate degree programs, 99, 118
Australia, case studies, 28
Austria, case studies, 28

bachelor's degrees, 122–126
barriers, to upgrading, 144
Bay State Skills Corporation, 113
blue collar workers, 5, 70–71, 84
bottlenecks: labor, 95; work flow, 45
branch plants: and high-technology
  employment, 153; and local
  economic development, 154,
  185n.12; and product life cycles,
  14, 166n.8

Canada, case studies, 28
career paths, 19–22, 55, 65, 71–72,
  143

case studies: annotated bibliography
  of, 203–221; data base 27–28;
  48–49; benefits of, 6–7, 45;
  limitations of 7, 44, 170n.1
CETA (Comprehensive Employment
  and Training Act), 113
clerical work: and deskilling, 36–37;
  and downgrading, 63; in high
  technology industries, 5, 84; and
  job enlargement, 39; and lateral
  transfers, 62; new positions, 43;
  and skill shortages, 94; and
  upgrading, 53; and widening
  skills gap, 71–72
colleges, 42, 60–61, 85, 98, 99, 117.
  See also community colleges,
  education and training
  institutions, higher education
community colleges: curricular mix,
  188; and the skill-training life
  cycle, 42, 92, 152; and vocational
  education funds, 116. *See also*
  education and training
  institutions
communities, and technological
  change 4–6, 8, 22, 144–146. *See also*
  local economic development
"company towns", 64
competitive advantage: in declining
  industries, 46; of educational
  institutions, 153; of high
  technology industries, 88–89; and
  local economic development,

ABOUT THE AUTHOR

Patricia M. Flynn is professor of economics at Bentley College where she also serves as executive director of the Institute for Research and Faculty Development. She received her Ph.D. in economics from Boston University.

Professor Flynn has been a research associate at the Institute for Employment Policy at Boston University and a senior research fellow at the New England Board of Higher Education. She has also served as faculty member of the Institute in Employment and Training Administration at Harvard University. In 1983–1984 she was a visiting scholar at the Federal Reserve Bank of Boston.

Professor Flynn's principal research interests are the impact of technological change on employment and economic development and the responsiveness of education and training systems to changing labor market needs. Her research has been supported by the U.S. Department of Labor, the National Institute of Education, and the Henry Rauch Faculty Enrichment Fund at Bentley College.

Professor Flynn is a member of the Executive Board of the American Council on Education's National Identification Program in Massachusetts, the Governor's Task Force on Higher Education and the Newton Economic Development Commission. She also serves on accreditation teams for the New England Association of Schools and Colleges (NEASC).